# Science and Technology Teacher Education in the Anthropocene

This unique book compares anthropogenic challenges in science and technology teacher education between the northern and southern contexts of Sweden and South Africa, respectively.

Presenting the results of a three-year research collaboration between science and technology teacher education researchers from South Africa and Sweden, the book explores theoretical perspectives and pedagogical experiences in response to challenges in the Anthropocene. It discusses research-informed practice in teacher education to address sustainable development. Chapters in the book collectively investigate the influence of current environmental and societal changes on the education of teachers, answering the question of how science and technology teacher education can adjust to current changes in the world and prepare new teachers for work in their future profession. Touching on issues such as climate change, global warming and pandemic diseases, the book uses a comparative approach and explores opportunities and possibilities for fulfilling the goals of science and technology education for sustainable development.

The book offers recommendations and opportunities to implement sustainability issues and develop sustainable teaching strategies. It will be a key reading for researchers, academics and post-graduate students in the fields of teacher education, science and technology education, sustainability education and comparative education.

**Miranda Rocksén** is Senior Lecturer at the Unit for Subject Matter Education with specialisation in Science and Technology Education, Department of Pedagogical, Curricular and Professional Studies, at the University of Gothenburg, Sweden.

**Elaosi Vhurumuku** is Associate Professor of Science Education and Chemistry Education specialist in the Division of Science Education, Wits School of Education, University of Witwatersrand, South Africa.

**Maria Svensson** is Associate Professor at the Department of Pedagogical, Curricular and Professional Studies, University of Gothenburg, Sweden.

**Emmanuel Mushayikwa** is Senior Lecturer in the Division of Science Education, Wits School of Education, University of Witwatersrand, South Africa.

**Audrey Msimanga** was Professor and Head of Wits School of Education, University of Witwatersrand, South Africa.

## Routledge Research in STEM Education

The *Routledge Research in STEM Education* series is home to cutting-edge, upper-level scholarly studies and edited collections covering STEM Education.

Considering science, technology, engineering, and mathematics, texts address a broad range of topics including pedagogy, curriculum, policy, teacher education, and the promotion of diversity within STEM programmes.

Titles offer dynamic interventions into established subjects and innovative studies on emerging topics.

**Teaching Early Algebra through Example-Based Problem Solving**
Insights from Chinese and U.S. Elementary Classrooms
*Meixia Ding*

**Gender Equity in STEM in Higher Education**
International Perspectives on Policy, Institutional Culture, and Individual Choice
*Edited by Hyun Kyoung Ro, Frank Fernandez, and Elizabeth Ramon*

**Mathematics Education for Sustainable Economic Growth and Job Creation**
*Edited by David Burghes and Jodie Hunter*

**Queering STEM Culture in US Higher Education**
Navigating Experiences of Exclusion in the Academy
*Edited by Kelly J. Cross, Stephanie Farrell, and Bryce Hughes*

**Science and Technology Teacher Education in the Anthropocene**
Addressing Challenges in the North and South
*Edited by Miranda Rocksén, Elaosi Vhurumuku, Maria Svensson, Emmanuel Mushayikwa and Audrey Msimanga*

For more information about this series, please visit: www.routledge.com/Routledge-Research-in-STEM-Education/book-series/RRSTEM

# Science and Technology Teacher Education in the Anthropocene

## Addressing Challenges in the North and South

**Edited by Miranda Rocksén, Elaosi Vhurumuku, Maria Svensson, Emmanuel Mushayikwa and Audrey Msimanga**

Routledge
Taylor & Francis Group
LONDON AND NEW YORK

First published 2022
by Routledge
4 Park Square, Milton Park, Abingdon, Oxon OX14 4RN

and by Routledge
605 Third Avenue, New York, NY 10158

*Routledge is an imprint of the Taylor & Francis Group, an informa business*



*British Library Cataloguing-in-Publication Data*
A catalogue record for this book is available from the British Library

*Library of Congress Cataloging-in-Publication Data*
A catalog record has been requested for this book

ISBN: 9781032039947 (hbk)
ISBN: 9781032039978 (pbk)
ISBN: 9781003190158 (ebk)

DOI: 10.4324/9781003190158

Typeset in Galliard
by KnowledgeWorks Global Ltd.

We would like to send a special thank you to Professor Audrey Msimanga who with her sharpness and humble manner has been an important person for the collaboration throughout the creation of this book. Your memory lives on through the texts you have contributed within the book.

# Contents

# Figures

# Tables

# Contributors

**Sakyiwaa Danso Boateng** holds a PhD in science education from the Wits University, Johannesburg. She is an early career researcher and currently a lecturer in science education at the Walter Sisulu University. Her research interests are on 21st century science teaching, science education and science teacher training.

**Git Börjesson** has been a teacher in science and technology school year 4–9, since 1994, and is now a teacher educator. She also takes part in a project on behalf of the National Agency for Education, developing an assessment material regarding science in school year 6.

**Anne-Marie Cederqvist** holds a PhD in technology education. The focus in her PhD project was teaching and learning programming in relation to technological solutions. She is a teacher in maths, science and technology for primary education, and at the University of Gothenburg, she is involved in courses focusing on technology education.

**Nina Christenson** works at Karlstad University in Sweden as a Senior Lecturer and holds a PhD in science education. She teaches at various teacher programmes and is head of an international research project focusing on how to promote climate literacy in school.

**Washington Dudu** research interests are in scientific inquiry, nature of scientific, pedagogical content knowledge and the affordances of indigenous knowledge in the science classroom. Washington has also been running the Experimento-Stiftung international project where he has trained a lot of practicing teachers in providing didactic and methodological approaches to classroom experiments.

**Anna Maria Hipkiss** studies the language-content relationship in school subjects. Her studies have illuminated a need for greater awareness of the importance of linguistic knowledge in teaching and learning of different school subjects. She currently works with developing the teacher education programme for primary education as regards language-content relationship.

**Portia Kavai** is a Life Sciences (Biology) teacher educator at the University of Witwatersrand in South Africa. Her research interest is in Life Sciences Education encompassing education for sustainable development (ESD), technological pedagogical and content knowledge (TPACK), STEM education and experiential learning.

**Malose Kola** has taught technology education for more than 20 years in different universities. The impact of technology on the environment has been one of the niche interests. Technology as a subject provides a platform for students to engage in real-life authentic solutions.

**Samuel Khoza** is a Senior Lecturer in the Technology and Vocational Education department at the Tshwane University of Technology. He facilitates undergraduate as well as postgraduate levels. He also supervises Masters and PhD students in the department. His research interests are in the Technical and Vocational Education and Training (TVET) field and beyond.

**Audrey Msimanga** is a Professor and was appointed as the Head of the School of Education in February 2021. Before that she served as the Head of the School of Education at Sol Plaatje University. Her research in science education focused on contexts of teacher and learner diversity and multilingualism. She was passionate about research and showed a great sense of discipline and academic wisdom.

**Emmanuel Mushayikwa** research interests include Science Education, Teacher Professional Development and Indigenous Knowledge Systems and Cultural Alignment of science education. Dr Mushayikwa is currently working on developing support for novice and out of field science teachers in rural schools in South Africa.

**Miranda Rocksén** is a Senior Lecturer at the University of Gothenburg and holds a PhD in Subject matter education specializing in the Natural Sciences. Her research interest include science teaching and learning processes across school levels with a special interest in the study of interaction.

**Suresh Kamar Singh** currently an Associate Professor in Science Education and the Head of Department for Mathematics, Science and Technology Education at the University of Limpopo. He is a teaching fellow who enjoys research, mentoring young academics and training the next generation of Physical Science teachers in South Africa.

**Marlene Sjöberg** research interest concerns student teachers' and teachers' development of professional knowledge and PCK regarding teaching and learning science, especially in the collegial setting. In addition, she is a teacher educator in science courses for student teachers and teachers at University of Gothenburg.

**Maria Svensson** is an Associate Professor in the subject area of technology at University of Gothenburg. Her research focus is in the field of technology education with a special interest in teaching and learning technological systems and system thinking.

**Maria Tsakeni** is a science teacher educator at the University of the Free State in South Africa. Her research interest is in science curriculum innovations that include education for sustainable development, STEM education, inquiry-based practical work and the integration of educational technologies in science teaching.

**Elaosi Vhurumuku** is a chemistry education specialist and Associate Professor of Science Education, in the School of Education, University of Witwatersrand. His research interests are in: laboratory work in science education, scientific inquiry, indigenous knowledge systems, education for sustainable development and teacher professional development.

**Ann-Marie von Otter** is a Lecturer of subject didactics at the University of Gothenburg. Her research interests concern teaching and learning in technology, the method Learning study, the use of digital tools and sustainable development in preschool, elementary school and higher education, with a focus on the development of teacher's didactic skills.

**Kassahun Weldemariam** is a Senior Lecturer at the Department of Pedagogical, Curricular and Professional Studies within the Faculty of Education at the University of Gothenburg in Sweden. Weldemariam's research interest falls under the umbrella subject of Education for Sustainable Development with a focus ranging from early years to higher education.

**Alexina Thorén Williams** is a teacher educator in science didactics at the University of Gothenburg in Sweden. Her research research interest is in student teachers' development of science didactics, integration of ESD and out-of-school science learning environments in teacher preparation and biomimicry education as a holistic approach to sustainable future.

**Sally Windsor** teaches in the international Masters of Education for Sustainable Development and teacher education programmes at Gothenburg University. Her interests include Indigenous knowledges and cultures for sustainability, school level sustainability education, geography teaching, professional conversations and practicum pedagogies. She is part of a collaborative European project "*Climateracy* – Developing European teachers' climate literacy", and in 2022 started a project titled "Growing through mentoring: An activity-based inquiry into mentor teachers' knowledge and practices".

# Preface

This book is the result of a research collaboration involving teacher educators and researchers from South Africa and Sweden. The ambition when we started the project was to investigate and develop our understanding of how we in science and technology teacher education courses prepare our student teachers to face the greatest challenges of our time, the increasing need to address sustainability issues. The work on the book has opened our eyes to different perspectives on how sustainable development can be expressed in teacher education but also the necessity of a change to achieve a paradigm shift in teacher education. Our argument is that this shift is necessary, if sustainable development is to be managed in this era of the Anthropocene. During the work with the book we also ended up in the middle of a global crisis with COVID-19. It has, in many ways, affected our work both in terms of the substantive parts where sustainability issues suddenly took on a different meaning, especially in terms of health and well-being, social and economic aspects and in terms of possibilities to meet and collaborate. Despite this or thanks to this, we perceive the content of the book as important and relevant, it is a contribution to a discussion about sustainable development and the changes that need to be made by many and in many ways for a continued life on this earth. The book provides useful insights for research and practice in science and technology teacher education.

*Miranda Rocksén, Elaosi Vhurumuku,*
*Maria Svensson and Emmanuel Mushayikwa*
*Johannesburg and Gothenburg*
*September 2021*

# Foreword

As the global sustainability challenges become more urgent by the day, the role of education in addressing these challenges receives more prominence as well. This is critical as education, certainly including science and technology education, in often subtle but significant ways, has laid the foundations for the current epoch we are in, the Anthropocene. After all, while mainstream science and technology education, rooted in the universities of the global North, through processes of colonialisation exported to the global South, has cultivated curiosity, knowledge and understanding, it also advanced particular ways of thinking and acting that we now are coming to see as one-sided and even dysfunctional in light of sustainability. These ways of thinking and acting include the desire and ability to breaking things down, creating order by making distinctions and drawing boundaries, maximising predictability and minimising uncertainty, striving for some kind of objectivity, stressing the cognitive while ignoring the socio-emotional and the embodied and always looking for possibilities to innovate, concur and expand. These ways of thinking and their accompanying "engineering mindsets" have brought many people across the world better lives in that they have access to all kinds of technologies that have made live a lot easier, although not necessarily more meaningful.

At the same time, the access to these technologies has not been equal; some people have more or better access than others. Illustrative for what might be called the dysfunctional technology push is the fact that more people across the globe have access to cell phones, than to clean drinking water, good health and sanitation (United Nations, 2013). Another troubling aspect of these new technologies and their global distribution is that they depend heavily on forms of "extractivism", varying from the mining of precious earth minerals and metals to the exploiting of people in the technology chain from the mine to the landfill or incinerator.

So, here is a book critically looking at science and technology education with the challenge of finding more sustainable ways of moving forward as a species in the coming decade. A book that seeks to blur boundaries, rather than to make distinctions. Questioning binaries, acknowledging "fluidity" and crossing boundaries might be a core task for science and technology education. In this book the North-South binary is overcome, at least to a degree, through collaboration between authors nested in different socio-economic and socio-cultural cultures in South Africa and Sweden, but all trapped in the same homogenising neo-liberal capitalist system that has made some scholars speak of the Capitalocene (blaming a destructive system for our current predicament) as an alternative to Anthropocene (blaming humans as one entity for our current predicament) (Moore, 2017).

I was privileged to be a part of the launch of this collaborative effort in 2018 where scholars and practitioners from South Africa and Sweden looked for common ground, a common way of speaking about science and technology education in light of SD (STESD) and crafted initial ideas for common themes to be explored from different vantage points. Now, three years later, the result is a rich collage of chapters seeking to provide a new narrative for science and technology education, one that provides a starting point for more generative and, indeed, more disruptive pathways that can transgress normalised patterns and routines in science and technology education that make education part of the problem and not part of the solution. The book offers clear alternatives from creating space for more embodied ways of teaching, critical digital literacy, more relational pedagogies that allow for boundary crossing between subjects, forms of participatory and design-based research seeking to overcome the research-practice binary.

2030, the year the UN Agenda and its 17 SDGs, and the goals of the Paris Agreement on Climate change need to be realised to increase the likelihood of homo sapiens surviving in a good way on this planet, is approaching fast. Only eight more years from the moment this book was published... The challenge for educators, school and university leaders, policymakers and all those affecting the education system is enormous. If there is one thing clear, also from reading this book, it is that "education-as-usual" is no longer an option. The nature of the sustainability crisis is such that only a major reorientation of education, science and technology education inclusive, will suffice. This book provides helpful insights and steppingstones for a much-needed transition in teaching, learning and capacity-building, one that hopefully leads to a world that is more sustainable than the one currently in prospect.

*Arjen Wals, Professor of Transformative Learning for Sustainability, Wageningen University*
*Romsdalen, Norway, October 1, 2021*

## References

Moore, J. W. (2017). The Capitalocene, part i: On the nature and origins of our ecological crisis. *The Journal of Peasant Studies*, *44*(3), 594–630. doi: 10.1080/03066150.2016.1235036

United Nations. (2013/March 21). *Deputy UN chief calls for urgent action to tackle global sanitation crises.* UN News. Accessed 01-10-2021 from https://news.un.org/en/story/2013/03/435102-deputy-un-chief-calls-urgent-action-tackle-global-sanitation-crisis#.UUwW_VrIxZ8

# Acknowledgements

The research on which this book is based is financially supported by the National Research Foundation (NRF) of South Africa; the Swedish Foundation for International Cooperation in Research and Higher Education (STINT); The Swedish Research Council [in Swedish Vetenskapsrådet, VR], and Swedish Research Council for Health, Working Life and Welfare (Forte), within the South Africa-Sweden Bilateral Scientific Research Cooperation Programme.

# A list of key terms

**Anthropocene** Although, the term Anthropocene is a much-contested term among environmental scientists and geologists, it refers to the time we are living in where the actions of people cause changes to the climate, land, water, environment, organisms and atmosphere. It represents a new world order where people are the driving force. The term Anthropocene is derived from Greek terms "anthropo" (human) and "cene" (new) and was coined by Nobel laureate Paul Crutzen. The action of people will dominate the environmental agenda into the future emphasising the need for education for sustainable development (ESD).

**Education for sustainable development (ESD)** Education for sustainable development (ESD) is a UNESCO initiative aimed at transforming education for sustainability. It is a broad initiative aiming at preparing learners of all ages to find solutions for the challenges of today. According to the UNESCO website:

> Education for Sustainable Development (ESD) empowers learners of all ages with the knowledge, skills, values and attitudes to address the interconnected global challenges we are facing, including climate change, environmental degradation, loss of biodiversity, poverty and inequality.
>
> Learning must prepare students and learners of all ages to find solutions for the challenges of today and the future. Education should be transformative and allow us to make informed decisions and take individual and collective action to change our societies and care for the planet.
>
> **(UNESCO, n.d.)**

**Sustainable development (SD)** Sustainable development is defined as the development that meets the needs of the present without compromising the ability of future generations (Glavič, 2020). At the same time what this means depends on who's needs that are used as the frame of reference. Sustainable development works an umbrella concept for different concerns about the environment (planet), society (people) and economy (profit), but is criticised due to not accounting for political and economic dimensions.

**Sustainability** In literature sustainability refers to the ability to sustain three pillars, namely, the environment (planet), society (people) and economy (profit). The concept is also criticised for being unspecific. Although there are different definitions for sustainability, to sustain means to provide support or to keep something going. Simply it implies people catering for their own needs without compromising the needs of the next generations. Sustainability is an important concept as people behave as if the planet has infinite resources whereas our resources are finite.

## References

Glavič, P. (2020). Identifying key issues of education for sustainable development. *Sustainability*, 12, 6500. doi: 10.3390/su12166500

UNESCO. (n.d.). *Education for Sustainable Development.* United Nations Educational, Scientific and Cultural Organization (UNESCO). https://en.unesco.org/themes/education-sustainable-development

# Introduction to the book

*Miranda Rocksén, Elaosi Vhurumuku, Maria Svensson, Emmanuel Mushayikwa and Audrey Msimanga*

## Introduction

This book is the result from a three-year collaboration that brought together science and technology teacher education researchers from the Universities of Gothenburg and Karlstad, Sweden and the Universities of Witwatersrand, Limpopo, Northwest and Sol-Plaatje, in South Africa in a series of research collaboration workshops, over a three-year period from 2019 to 2021. Driven by the agenda to promote science and technology education for sustainable development (STESD), the workshops focused on comparing experiences, theoretical perspectives and sharing research findings as well as working collaboratively on research publications in response to current challenges.

The reality of the present epoch – the Anthropocene – is that humans through various activities including industrialisation have significantly altered the earth. Some would argue that the alleged abuse and irresponsible use of science and technology in the profit-driven corporate realm has given rise to a mesh of problems and challenges. These include such issues as global warming, habitat loss, and new diseases. Indeed, scientific debates have been forwarded challenging the realities and existence of many issues. Sadly, many of these arguments are used to justify continued corporate exploitation of energy and natural resources or the application of new technology. One example is the fracking debate in the Karoo in South Africa. Other examples are found in mining, forestry and genetic technology. In society at large these debates have blurred the distinction between scientific fact and fallacy, giving rise to post-truth conspiracies (see, for example, Feinstein & Waddington, 2020). They are nurtured by debates about the limitations of scientific knowledge and have roots in scientific ignorance, emotions, beliefs and political and economic agendas shared among groups such as climate change denialists, who are adversaries to the pursuit for sustainability. A growing number of societal issues with a scientific basis are sometimes included in science teaching and learning in the form of controversial socio-scientific issues (e.g., Kolstø, 2001). As Brennan (2017) observes, difficulties in debates about what it is that teacher education can do to contribute towards resolving sustainability challenges arise because these are intricately intertwined with the crises of capitalism. As education researchers and practitioners, we are in a quandary

DOI: 10.4324/9781003190158-1

of what it is that teacher education can do to contribute towards sustainable development, especially since in the institutional structures we are working in, interdisciplinary and integrated curriculum approaches have largely disappeared (Brennan, 2019). Brennan (2017) argues for pragmatism and a rethinking of the approaches and strategies in teacher education for environmental sustainability. In this book we contribute to debate around these pertinent issues.

Part of our effort in this book is to examine what the problems and challenges of the Anthropocene mean for science and technology teacher education; and what can be done and how teacher education may be reformed and reinvented to achieve the goals and aspirations of sustainability. While we do this, we are not under any illusions that teacher education is the panacea to the problems and challenges of the Anthropocene. However, we contend that science and technology teacher education can make meaningful contributions to the understanding of sustainable development. It is our argument that, in this era of the Anthropocene, to develop a discerning citizenry, it is necessary to build a high-quality Science and Technology Teacher Education (STTE), which can be accomplished not only through educational reform, but also through international research collaborations and partnerships.

## *About this book*

> A plastic bag is made from petroleum and will last for up to 100 years if it is deposited. Only in EU, citizens use almost 100 billion bags per year. Globally plastic constitutes around 10% of the total weight of municipal waste, which has made plastic pollution of our oceans a serious problem (*Steensgaard et al., 2017*). The pupils in today's schools will in 10–20 years be grown-ups making their own living, their own decisions and their own future. What knowledge and competencies do we want to pass on to them about the raw materials used for production, the manufacturing process, the lifecycle, energy, effects on atmospheric composition and alternative materials? What role do we think that science and technology education have, and do we consider this important? With global challenges we think that questions such as these are increasingly a matter to be discussed and responded to within the international community.

The challenges of the Anthropocene require that universities and other teacher education institutions make STESD and Global Citizenship Education (GCED) an integral part of their teacher preparation programmes. Karaarslan and Teksöz (2016) argue that it is not enough to just integrate STESD into the curricula of teacher preparation programmes, but also important to ensure that the teachers have adequate knowledge, competences and skills, enabling them to develop learners into citizens who understand and appreciate the environmental, social, economic and political aspects of sustainability. As noted, these aspects cover such issues as global warming, climate change, processes of migration, consumer literacy, democratisation, health, HIV/AIDS, biodiversity, habitat loss, species extinction, constructive citizenship and changes in the chemical composition of the atmosphere. For both the developed North and the developing South there is an understanding that integration of STESD into education programmes can contribute to teachers' understanding of United Nations' SDGs and achieving the goals of science education for scientific and sustainability literacy (see, for example, Bourn, Hunt, & Bamber, 2017; Laurie, Nonoyama-Tarumi, Mckeown, & Hopkins, 2016). However, there is an underlying assumption that, given the economic disparities between the North and the South and the differing demands on the educational systems, STESD could be prioritised, approached to and implemented in different ways in the North and the South. Given this, the interest of our comparative study was to compare how a country in the North (Sweden) and a country in the South (South Africa) approach and implement STESD in teacher education. The project was premised on the conviction that bringing together a team of multidisciplinary researchers and teacher educators from the two countries could go a long way towards contributing to crafting and improvising solutions that can help the practice of STESD. The book is the result of a research collaboration with a common interest in sustainability. There is hope that the findings disseminated in this book can contribute towards sensitising policymakers, curriculum developers and implementers about the importance of STESD as a critical aspect of the Anthropocene.

National systems of education are increasingly challenged to develop, collaborate and learn in relation to each other, in a rapidly changing global environment. Questions of sustainability, climate change, processes of migration and democratisation emphasise the importance of developing high-quality education – not only through educational reforms, but also through international research collaboration and partnerships. In particular, partnerships between institutions offering initial teacher education programmes have the potential to contribute significantly to educational improvements. In this chapter, we introduce the main themes of this book, and recapitulate the work process we have been involved in, over a period of more than three years.

The chapters and their topics as well as the outline of the book were the result from joint discussions during three workshops, two done in Sweden, and one in South Africa. By repeated exchanges, discussions and meetings – including online via ZOOM or SKYPE – the understanding and knowledge of what sustainability issues might mean in our two parts of the world, in relation to the education of science and technology teachers, have eventually grown and deepened. The work was guided by the overarching question: *how may science and technology teacher educations, adjust to current changes in society and the world and prepare new teachers appropriately for work in their future profession?* Within the broader theme of how to respond to such a broad and complex question, for each chapter, a group of authors representing the North and the South jointly developed a text on a specific topic. Each chapter can be read as a stand-alone text, with its own defined questions, methods of investigation and conclusions.

The book takes the form of a comparative case study conducted to respond to four major aims:

1 To critically appraise some of the theoretical perspectives that have informed and guided science and technology teacher education in the Anthropocene;
2 To compare policies and curricula in university science and technology teacher education in South Africa and Sweden;
3 To investigate and compare experiences and pedagogical practices in science and technology teacher education in South Africa and Sweden; and
4 To discuss opportunities and possibilities for fulfilling the goals of science and technology teacher ESD, in the light of anthropogenic challenges.

As mentioned, this book is the result of a collaboration between science and technology researchers and teacher educators at universities in South Africa and Sweden. South Africa and Sweden are in the northern and southern hemispheres, respectively. Both countries are within the vicinity of the earth's corresponding two poles. With a population of 59 million, South Africa has nearly six times as many inhabitants compared to Sweden's 10 million people. This makes the population density of South Africa 49 persons per square-kilometer, which is about twice the population density of Sweden, 25 persons per square-kilometer (Food and Agriculture Organization and World Bank, 2020). South Africa has a Gross Domestic Product (GDP) per capita of $5 090, while for Sweden, the GDP per capita is $51 925 (The World Bank, 2020). On the Human Development Index (HDI), South Africa is pegged at 0.739 while Sweden has an HDI of 0.945 (UNDP, Human Development Report, 2020). These indicators point towards disparities between the two countries in terms of the levels of basic dimensions of human development: health, education and living standards. Thus, while the two countries share common interests in promoting sustainability, the disparities inherently necessitate differentials in approaches and strategies in tackling anthropogenic problems and challenges. Indeed, this reality is an unsurprising finding from the research presented in this book.

The work with the book was done in a project, that, in order to relate to current times, borrowed in its title the word Anthropocene. The project title is Sweden-South Africa universities research collaboration: science and technology initial teacher education in the Anthropocene. The term Anthropocene was first proposed as a potential geochronological unit in the year 2000 (Crutzen & Stoermer, 2000). In 2017, the Anthropocene Working Group of geologists decided to propose the term to be formalised as an epoch, based on the influence of humans on key physical, chemical and biological processes at a planetary scale (Zalasiewicz et al., 2017). The formal process involves international geological organisations collecting evidence and deciding on possible borders of this proposed epoch. The word *anthropos* is Greek for human, and the word Anthropocene has come to represent more than a geological epoch. It is a word that acknowledges that environmental problems have grown so big that one considers human impact to drive this far-reaching and rapid change at a planetary scale (Angus, 2015; Corlett, 2015). The scale of the changes – in climate, biodiversity, oceanic acidification, atmospheric composition and more – is brutally illustrated by the measures of control variables in the model of nine planetary boundaries (Steffen et al., 2015).

In research, the word Anthropocene has come to represent theoretical perspectives of our time, through which researchers are critically analysing, for example, human's dominant position with regard to other biological organisms or the non-living world. In educational research those perspectives invite rethinking the role of education and discussing the relation between the concept and education more generally (Bonnett, 2019). The connection of Anthropocene to the fields of geology, critical theory and education, makes it a multifarious, disparate and convoluted concept. It is a construct that can be understood in an overabundance of ways, based on theoretical, paradigmatic and ideological tastes and orientations. In the work with this book, we have chosen to use Anthropocene as an overall frame that enables us to critically analyse our own respective, policies, practices and curricula with fresh eyes in light of the contemporary. We do not want the theoretical concept to overshadow the different examples, data comparisons, results, as well as the different perspectives of what the United Nations' SDGs (can) mean in relation to the science and technology teacher education programmes that we present. For this purpose, we are using/sharing the definition of Anthropocene provided by the geologists, with one of the chapters (Chapter 3), elaborating more on the concept. As a whole, the book can be seen as a manifestation of North and South perspectives on science and technology teacher education in our time.

From our two perspectives –the North and the South – we together investigated influences that current environmental and societal changes already have on the education of teachers. On the global scale, the 2030 Agenda for Sustainable Development adopted by all United Nations Member States in 2015 has greatly influenced discussions in education (see, for example, Agbedahin, 2019; González García, Colomo Magaña, & Cívico Ariza, 2020). This document has been useful as one common reference, with the 17 SDGs at its heart. As noted, today, becoming a teacher in science and technology means to develop competences to be able to teach about subject-specific concepts and principles, but also to develop competences to teach about interdisciplinary topics, as well as discuss the social, economic, ecological, political and ethical aspects pursued by questions of sustainability. The increasing availability of information technology has an influence on the teaching practices and resources being used. Heterogeneous student groups point to the importance of learning to consider the subject-specific resources that a multiplicity of languages and cultures make available in schools. Learner differentiation has become an important factor in teaching and learning. The book chapters reflect some of this complexity.

In both countries, education is today compulsory up to the age of 15, after which secondary education is optional, and may include specialisation in the natural sciences, technology or engineering. To educate certified teachers, universities and other high education institutions provide initial teacher education for primary and secondary education teachers. In South Africa and Sweden, education has experienced reforms. In South Africa, the new CAPS documents in science education replaced previous outcome-based initiatives (Department of Basic Education, 2012). In Sweden, a decentralised governance of education at primary and secondary levels has induced several national reforms, for example revisions of curricula, national testing systems and the development of a national school inspection (Swedish National Agency for Education, 2010). In science and technology education questions concerning sustainability and citizenship are highlighted and included in the school curriculum, which provides opportunities for creating links in the contexts of educational practices (Wals, Brody, Dillon, & Stevenson, 2014). In higher education, synergy between different fields of research points towards Responsible Research and Innovation, RRI (Tassone, O'Mahony, McKenna, Eppink, & Wals, 2018), and projects like PARRISE[1] which aim to support science teachers at the levels of primary and secondary school to include socio-scientific issues in their teaching. Significant in teacher education in Sweden, is for example, developing research-informed practices covering subject-specific teacher competence, breadth in instructional strategies and how to address sustainable development issues. In South Africa, noteworthy in teacher education, has been, for example, developing teaching approaches that integrate indigenous knowledge systems and culturally embodied learning. Together with what is significant in each country there are an increasing number of new and important issues of concern at the regional and global levels that must be considered in teacher education. During the years 2020 and 2021 the COVID-19 pandemic is the latest example of a crisis that dramatically has affected education globally, but also in our two countries, which necessitated closing schools and remote teaching.

The data sources for the articles in this book are documents, interviews and classroom observations. We make use of a methodology for comparative and international studies in education described by Phillips and Schweisfurth (2014). We used the described model to organise our study of the data. The model provides us with a systematic way to approach a new and difficult area. It enabled us to narrow down a broad and common interest and identify areas that required our investigation and how to illuminate the significant details. Following the footsteps of Phillips and Schweisfurth (2014), our comparative inquiry involves six stages (Figure 0.1). The first stage *Conceptualisation* means that the questions to be addressed are formulated as neutral from any particular context. The second stage, *Contextualisation*, follows, with a detailed description of the educational phenomena against the local background. At this stage the investigation includes details of those historical, geographical, cultural, political and linguistic aspects that make a characteristic of the local context. The third stage, *Isolation*, means to make a direct comparison, to isolate differences and conduct an analysis of the variables. *Explanation* is a fourth stage in which hypothesis may be formulated, which may explain the observed differences in the two contexts. The fifth and final stage involves *Application* and takes the findings to other contexts testing their applicability and if possible, generalise. In the book the stage of

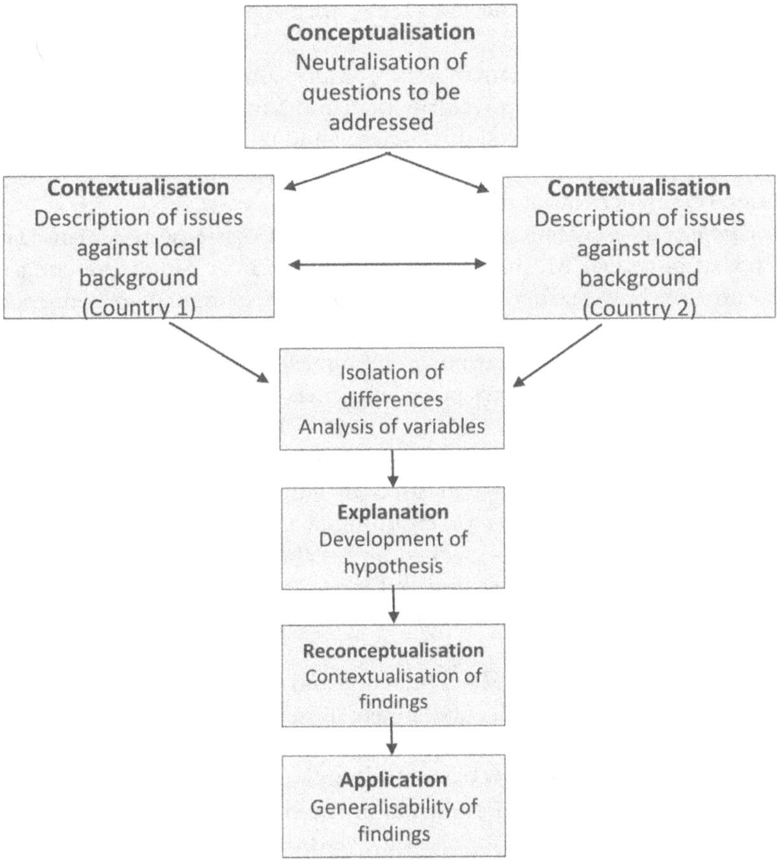


*Figure 0.1* A structure for comparative inquiry from Phillips and Schweisfurth (2014, p. 119)

Contextualisation is represented by Chapters 1–9. Each chapter defines one or two questions specific to the investigated area (conceptualisation), which is then contextualised by description of the cases. Even if both countries, South Africa and Sweden, are accounted for in every chapter, the balance between them may vary. Each chapter ends by a conclusion section in which the differences and similarities found are summarised, the Isolation.

The book consists of two parts. Part I has four chapters and includes theoretical perspectives on issues of sustainability in relation to science and technology teacher education. Part II has six chapters that investigate different aspects of the practices of teacher education through curricula analyses, interviews, classroom observations and surveys.

In *Chapter 1*, Miranda Rocksén, Elaosi Vhurumuku and Maria Svensson present a literature review on how sustainability is addressed in research on science and technology teacher education, focusing on studies done in South Africa and Sweden.

*Chapter 2*, by Suresh Kamar Singh, Anna Maria Hipkiss, Portia Kavai, Washington Dudu, Kassahun Weldemariam and Nina Christenson, presents and analyses the policies and curricula in science and technology education, in relation to sustainability, focusing on the two participating countries, with special attention on teacher education.

In *Chapter 3*, Kassahun Weldemariam, Nina Christenson, Alexina Thorén Williams, Emmanuel Mushayikwa and Suresh Kamar Singh present a critical inquiry on how teacher education programmes respond to contemporary anthropogenic challenges. Their inquiry is carried out through a comparative analysis of four teacher education programmes in South Africa and Sweden, looking at how the programmes prepare teachers to engage with anthropogenic challenges. Theoretically, the study is framed within critical inquiry framework.

*Chapter 4*, by Audrey Msimanga, Miranda Rocksén and Anna Maria Hipkiss, discusses the role of teacher education in addressing the challenges of the Anthropocene in the context of multilingualism. Using authentic examples from science classrooms in the two countries they illustrate the many semiotic resources at play in science teaching and learning, arguing that science teachers need to learn how to make use of these resources in developing scientific and sustainability literacy.

In *Chapter 5*, Malose Kola, Maria Svensson and Ann-Marie von Otter tackle the question of the extent to which environmental issues are part of the design process in technology education. They do this using two case studies, investigating examples of how the design process is used in technology education in South Africa and Sweden, in relation to the environmental situation in the Anthropocene.

*Chapter 6* by Sally Windsor, Maria Svensson, Washington T. Dudu and Marlene Sjöberg investigates the role of teaching practice in the education of science teachers. They highlight what professional knowledge and practices student teachers say are critical for practice in Science, Technology, Engineering and Mathematics (STEM) in this age of the "Anthropocene". Their findings reveal what teacher educators and student teachers from two teacher education programmes in South Africa and Sweden say is needed for current science, technology and sustainability education.

In *Chapter 7*, Portia Kavai, Alexina T. Williams and Maria Tsakeni present a comparative study that explores how South African and Swedish teacher education programmes prepare prospective Biology teachers for ESD. Their findings provide insights into the tensions and challenges that are navigated by teacher educators. They recommend expanded research into ESD in teacher education.

In *Chapter 8*, Emmanuel Mushayikwa, Miranda Rocksén, Sakyiwaa Danso Boateng, Samuel Khoza and Git Börjesson investigate two teaching interventions using the concept "culturally embodied learning" as an interpretive frame. The interventions involve activities in relation to teaching and learning of two areas in physics. In South Africa they interviewed teachers and learners about how electricity and magnetism were used to develop scientific and sustainability literacy.

In Sweden, the teaching and learning of force and motion were investigated as an activity with student teachers at a children's playground.

In *Chapter 9*, Anne-Marie Cederqvist, Ann-Marie von Otter, Portia Kavai and Samuel Khoza present the results from a study focusing on the relationship between ESD and digital competence in the experiences of science and technology teacher educators in South Africa and Sweden. They analyse and compare the curricula of the two countries to establish similarities and differences that could inform the discussion on the extent to which the involved teacher education institutions prepare science and technology teachers for teaching digital competence with respect to ESD.

*Chapter 10*, by Elaosi Vhurumuku, Miranda Rocksén, Maria Svensson, Emmanuel Mushayikwa and Audrey Msimanga, gives a summary of the findings from the book project as well as insights for future research. Issues that connect local, regional and global questions in ESD and EE are fleshed out. A perspective of the Anthropocene is posited. Future challenges for science and technology teacher education programmes in the Anthropocene are teased out and interrogated.

## Note

1. https://www.parrise.eu/.

## References


Agbedahin, A. V. (2019). Sustainable development, education for sustainable development, and the 2030 agenda for sustainable development: Emergence, efficacy, eminence, and future. *Sustainable Development*, *27*(4), 669–680.

Angus, I. (2015). When did the Anthropocene begin… and why does it matter? *Monthly Review*, *67*(4), 1.

Bonnett, M. (2019). Interpreting the idea of the Anthropocene, and its relevance to education. *On Education. Journal for Research and Debate*, *2*(4). https://doi.org/10.17899/on_ed.2019.4.2

Bourn, D., Hunt, F., & Bamber, P. (2017). A review of education for sustainable development and global citizenship education in teacher education. (UNESCO Global Education Monitoring Report). London: UCL Institute of Education.

Brennan, M. (2017). Struggles for teacher education in the age of the Anthropocene. *Journal of Education (University of KwaZulu-Natal)*, (69), 43–66. Retrieved January 16, 2022, from http://www.scielo.org.za/scielo.php?script=sci_arttext&pid=S2520-98682017000200003&lng=en&tlng=en

Brennan, M. (2019). Changing teaching and teacher education in the 'Anthropocene'. *Journal of Research and Debate*, *2*(4), 1–6.

Corlett, R. T. (2015). The Anthropocene concept in ecology and conservation. *Trends in Ecology & Evolution*, *30*(1), 36–41.

Crutzen, P. J., & Stoermer, E. F. (2000). The Anthropocene. *IGBP newsletter, 41.* Stockholm, Sweden:Royal Swedish Academy of Sciences. http://www.igbp.net/publications/globalchangemagazine/globalchangemagazine/globalchangenewsletter-sno4159.5.5831d9ad13275d51c098000309.htm.l

Department of Basic Education. (2012). *Curriculum Assessment Policy Statements (CAPS)*. South Africa: Republic of South Africa. Retrieved from http://www.education.gov.za/Curriculum/CurriculumAssessmentPolicyStatements%28CAPS%29/CAPSSenior.aspx

Feinstein, N. W., & Waddington, D. I. (2020). Individual truth judgments or purposeful, collective sensemaking? Rethinking science education's response to the post-truth era. *Educational Psychologist*, *55*(3), 155–166.

González García, E., Colomo Magaña, E., & Cívico Ariza, A. (2020). Quality education as a sustainable development goal in the context of 2030 agenda: Bibliometric approach. *Sustainability*, *12*(15), 5884.

Karaarslan, G., & Teksöz, G. (2016). Integrating sustainable development concept into science education program is not enough: We need competent science teachers for education for sustainable development – Turkish experience. *International Journal of Environmental and Science Education*, *11*(15), 8403–8425.

Kolstø, S. D. (2001). Scientific literacy for citizenship: Tools for dealing with the science dimension of controversial socioscientific issues. *Science Education*, *85*(3), 291–310. doi: https://doi.org/10.1002/sce.1011

Laurie, R., Nonoyama-Tarumi, Y., Mckeown, R., & Hopkins, C. (2016). Contributions of education for sustainable development (ESD) to quality education: A synthesis of research. *Journal of Education for Sustainable Development*, *10*(2), 226–242.

Phillips, D., & Schweisfurth, M. (2014). *Comparative and international education: An introduction to theory, method, and practice*. London: Bloomsbury.

Steensgaard, I. M., Syberg, K., Rist, S., Hartmann, N. B., Boldrin, A., & Hansen, S. F. (2017). From macro-to microplastics: Analysis of EU regulation along the life cycle of plastic bags. *Environmental Pollution*, *224*, 289–299. https://doi.org/10.1016/j.envpol.2017.02.007

Steffen, W., Richardson, K., Rockström, J., Cornell, S. E., Fetzer, I., Bennett, E. M., & Sörlin, S. (2015). Planetary boundaries: Guiding human development on a changing planet. *Science*, *347*(6223), 1259855. https://doi.org/10.1126/science.1259855

Swedish National Agency for Education. (2010). *A presentation: The Swedish National Agency for Education supervises and provides support for better schooling*. Retrieved from https://www.skolverket.se/publikationer?id=2492

Tassone, V. C., O'Mahony, C., McKenna, E., Eppink, H. J., & Wals, A. E. (2018). (Re-) designing higher education curricula in times of systemic dysfunction: A responsible research and innovation perspective. *Higher Education*, *76*(2), 337–352.

The World Bank. (2020). Population density (people per sq. km of land area). *World Development Indicators*. Retrieved September 26, 2021, from https://data.worldbank.org/indicator/EN.POP.DNST

The World Bank. (2020). GDP per capita (current US$). *World Bank national accounts data, and OECD National Accounts data files*. Retrieved September 26, 2021, from https://data.worldbank.org/indicator/NY.GDP.PCAP.CD?locations=ZA&name_desc=false

United Nations Development Programme. (2020). *Human development report 2020: The Next Frontier: Human Development and the Anthropocene*. Retrieved September 26, 2021, from http://hdr.undp.org/en/content/latest-human-development-index-ranking

Wals, A. E., Brody, M., Dillon, J., & Stevenson, R. B. (2014). Convergence between science and environmental education. *Science*, *344*(6184), 583–584.

Zalasiewicz, J., Waters, C. N., Summerhayes, C. P., Wolfe, A. P., Barnosky, A. D., Cearreta, A. … Williams, M. (2017). The working group on the Anthropocene: Summary of evidence and interim recommendations. *Anthropocene*, *19*, 55–60. https://doi.org/10.1016/j.ancene.2017.09.001

# Part I

# Theoretical Perspectives, Policies and Curricula in Science and Technology Teacher Education

# 1 Sustainability and science and technology teacher education

## A review of the literature

*Miranda Rocksén, Elaosi Vhurumuku and Maria Svensson*

## Introduction

In the contemporary era of the Anthropocene, globally, there is increasing consensus that the use of energy and natural resources is no longer sustainable. There is a frightening and alarming realisation that humanity is recklessly pushing the earth beyond the limit necessary for preserving planetary ecological systems, galloping the earth towards cataclysm (Hamilton, 2017; Sepkoski, 2020). While this is so, there is some measure of universal agreement that research-based science and technology teacher education can make important contributions to sustainability or sustainable development (Eilks, 2015). Nevertheless, it appears that there is no single answer to the question of what a research-based science and technology teacher education means.

In many countries, politicians have generally agreed that the challenges of the Anthropocene require that teacher education programmes integrate science and technology education for sustainable development (ESD) into their curricula (Fuertes-Camacho, Graell-Martín, Fuentes-Loss, & Balaguer-Fàbregas, 2019). There is an underlying assumption that this will contribute towards achievement of the United Nations' Sustainable Development Goals (UN SDGs) through developing teachers' scientific and technological literacy, emphasising the importance of fighting for global sustainable development. It is contended that developing teachers' scientific and technological literacy is not only related to, but a necessity, for a more sustainable future. Nonetheless, in order to prepare student teachers properly, it is not only necessary to include science and technology ESD in curricula, but also essential to glean insights from critical literature perspectives and research results.

It suffices to mention that the research literature on sustainable development in teacher education is not one articulate field. Many of the visions in the UN SDGs have been expressed in such research fields as ESD and environmental education (EE). These efforts aboard, disparate and convoluted views on what necessary skills student teachers need in order to contribute meaningfully to sustainable development efforts. Moreover, debates around these issues are based on varying theoretical perspectives and ideological orientations (Knutsson, 2013). A common feature threading these discussions is the acceptance that, through

DOI: 10.4324/9781003190158-3

ESD, student teachers are expected to develop many complex competencies (Wiek, Withycombe, & Redman, 2011) necessary for participation in sustainable development. In school practice, ESD requires the teacher to embrace and cognise a different understanding of both teaching and learning (Dahl Madsen, 2013). The research field is also permeated by the policy statements, stakeholder expectations as well as views expressed by various interest groups, both at the national and international arenas. In 1972 the United Nations' Conference on Human Environment was held in Stockholm, Sweden. In 1982 *sustainable development* was established and defined as a concept by the report, Our Common Future (World Commission on Environment and Development, 1987) and in 2002 the World Summit on Sustainable Development was held in Johannesburg, South Africa. By the year 2021 the Cop26 conference on climate change was held in Glasgow. These events coexist with research on the global arena and can be expected to have strong influence on the direction of research. As alluded to, in many parts of the world the achievement of UN SDGs is an important goal for education in general and science and technology teacher education in particular.

In line with the theme of this book, in this chapter, we delve into the issue of how sustainable development has been approached focusing on the research literature in science and technology teacher education. Starting in the year 1980, just before the Brundtland report (World Commission on Environment and Development, 1987) and ending in 2020, we look at 40 years of publications in the two countries, South Africa and Sweden. Focusing on these two countries, our aim is to explore how environmental issues and sustainability have been conceptualised in relation to teacher education and the fields of science and technology, covering the period from 1980 to 2020. We start with a presentation of some conceptualisations of the ESD and EE research fields. This is followed by an examination of a sample of pertinent research literature from South Africa and Sweden. Within this effort, we provide some useful insights for the practice of science and teacher education, for both the global North and the South.

## Figuring the field

### *Eilks' four models of integrating ESD into science education practices*

From the perspective of science education, Eilks (2015) reviews central frameworks of ESD and presents four models for connecting ESD and science education. Model 1 means adopting principles from sustainable practices in science and technology for hands-on science education laboratory work. This entails minimising the use of harmful chemicals and performing small-scale experiments. Students' learning can be expanded by their critical reflection and comparison of these strategies. In model 1, this idea of an ESD school laboratory mirrors the shift towards industrial "green" production. Model 2 means adding sustainable science as content, in science and technology curricula. Here, sustainable science and technology are topics addressed in the teaching. For instance, energy efficient processes, innovative products and new methods for products based on renewable resources. By focusing only on the technological applications, Eilks argues that there is a risk that this model does not allow students to get a thorough understanding of the interplay between science, technology or society, which means the interplay of economic, ecological and societal sustainability. Model 3 means to use controversial sustainability questions for the socioscientific issues driven science education. When using model 3 science education most often does not focus on neither the learning of science nor the learning about sustainability issues. For students the topics that are presented and interrogated imply that they need to include "both scientific knowledge and reflection upon societal debates on the practical, technical applications of such knowledge as factors to be learned" (Eilks, 2015, p. 155). In model 4, science and technology education is integrated into a development of a whole institution, like a school, along the main lines of ESD. Eilks describes this fourth model as an open classroom in which pupils and teachers engage in non-formal school projects, with visits to university, industry or research centres. All involved are required to explore future challenges and reflect on ESD-learning and their active participation in society. The four models presented by Eilks (2015) represent a synthesis of 15 years of integrating ESD into science teaching and science education. Other syntheses provide ideas about the historical development of research in ESD, in which the number of publications today is growing exponentially (e.g. Pavlova, 2013; Zhang & Wang, 2021).

## *Fifteen currents in environmental education*

As research turfs, the ESD and EE fields are broad and not easily digested. In the 1960s the book *Silent Spring* (Carson, 2002) had a big influence and EE was introduced in many Western countries, including Sweden. The book worked as an eye-opener concerning the effects of pesticides on wildlife and environment. Today, environmental problems have grown so big that it is globally accepted that the earth has entered a new epoch (Zalasiewicz et al., 2017) called the Anthropocene. Over the years, for both EE and ESD, theoretical perspectives and the foci of research have fluctuated. For example, Sauvé (2005) recognises the lack of consensus and describes 15 currents in EE (see Figure 1.1). These 15 currents are separated into those with a historically longer tradition and those that more recently emerged.

Currents with a longer tradition in EE are calling for action. This has entailed calling for changes in individual behaviour or collective action. These currents approach EE from a cognitive perspective. For example, one of the currents with a longer tradition is *Problem-solving*, where the goal is to inform or help people to inform themselves and learn about environmental issues, as well as to develop attitudes and skills for solving them. Another example is the *Systemic* current that allows for identification of the various components of an environmental situation or issue. Currents that more recently emerged in EE are connected to

| Among those Currents with a Longer Tradition in Environmental Education | Among those Currents more Recently Emerged in Environmental Education |
|---|---|
| 1. Naturalist Current | 8. Holistic Current |
| 2. Conservationist/Resourcist Current | 9. Bioregionalist Current |
| 3. Problem-Solving Current | 10. Praxis Current |
| 4. Systemic Current | 11. Socially Critical Current |
| 5. Scientific Current | 12. Feminist Current |
| 6. Humanist/Mesological Current | 13. Ethnographic Current |
| 7. Value-centred Current | 14. Eco-Education Current |
| | 15. Sustainable Development/Sustainability Current |

*Figure 1.1* Fifteen currents in environmental education, from Sauvé (2005)

such theoretical perspectives as critical theory and socio-ecological theory in combination with globality. One example is the *Holistic* current in which the global perspective is as central as the to the importance of embodiment in our relations with the environment, allowing beings (plants, animals, rocks, etc.) to speak by themselves about their own nature and in that way enable us to better care for them. The kind of analysis proposed by Sauvé (2005) allows for the construction of a "bigger picture" view, which enables a synthesis of the reality under study. In the line-up of currents in EE presented by Sauvé, the 15th current is called "Sustainable Development/Sustainability Current". Sauvé (2005, p. 30) writes:

> According to the supporters of this current, environmental education has limited itself to a naturalist approach and has neglected to encompass social preoccupations, and especially economic considerations, in the treatment of environmental questions. Education for sustainable development would permit that deficiency to be mended at last.
>
> (Sauvé, 2005, p. 30)

This highlights the interconnectedness between *economic*, *social* and *environmental* aspects of sustainable development, which have been called the three pillars of sustainability (Purvis, Mao, & Robinson, 2019).

### *Three or four pillars of sustainable development*

The three pillars were presented in 2001 as a "common view" of sustainable development (Giddings, Hopwood, & O'Brien, 2002). However, there are no clearly substantiated theories, forming a basis for the development of this framework. According to Purvis et al. (2019), the three pillars are used either as a system approach or as a perspective of sustainability. The system approach typically presents three distinct systems (*economic*, *social* and *environmental*) each with its own "goals". It is posited that the interactions among these systems must be managed to meet both the goals of each system and holistically the emergent goal of sustainability or sustainable development. Perceiving the three pillars as distinct perspectives of sustainability involves positioning the perspectives in opposition to each other. It is noteworthy that, within this realm, the meaning of the economic pillar remains a central point from which much of the early literature diverges. The conceptualisation of three pillars of sustainable development has been subjected to critique (see, for example, Colantonio, 2009; Lehtonen, 2004). Although the economic, environmental and social aspects are closely interconnected in theory, more recent literature reveals that both integrated and separated views on the three dimensions of SD are represented among upper secondary students (Berglund & Gericke, 2016). Sandell and Öhman (2010) suggest that a fourth dimension to be added: *sustainable ethics*. In describing this fourth dimension, they emphasise the role of experience with nature. In their view, sustainable ethics "…is rather a comprehensive existential perspective that originates from aesthetic and emotional relations with nature – the direct encounter with nature …" (Sandell & Öhman, 2010, p. 125).

### *Paradigmatic changes of teaching practices*

More recent compilations of the development of the EE and ESD fields show the "new currents" emerging after 2005 relating to post-humanism and new materialism (see Alaimo, 2012; Varela-Losada, Vega-Marcote, Pérez-Rodríguez, & Álvarez-Lires, 2016; Wals, Weakland, & Corcoran, 2017). For example, the review by Varela-Losada et al. (2016) takes the literature on *action competence* (see Jensen & Schnack, 1997) as a starting point when they study educational proposals in formal ESD. They claim the need for a paradigm promoting a complex and critical view of ESD, a paradigm that implies changing activities and relations in the teaching practice:

> This approach implies a teaching and learning model based on social learning (Lave & Wenger, 1991; Wals, 2007), that implies new student (and teacher) roles, where their interests and needs are taken into account (Barratt-Hacking, Barratt, & Scott, 2007; Mogensen & Schnack, 2010). If we understand the environment as a system made up of physical, socio-cultural and emotional factors that are all interrelated, it is essential to use the EE models that provide a holistic view of the world and of the environment. These real-world situations should be integrative at a global or interdisciplinary level (Mogensen & Mayer, 2005; Wals, 2007), and involve student participation in the classroom and in the resolution of environmental issues (McCallum, Hargreaves, & Gipps, 2000; Mogensen & Schnack, 2010). All this needs to be carried out using critical thinking, embracing complexity and studying

> future alternatives (Mogensen & Mayer, 2005; Wals, 2007; Kyburtz-Graber, 2013), encouraging autonomous and informed decision-making through participation (Wals, 2007; Stevenson & Stirling, 2010).
>
> (Varela-Losada et al., 2016, pp. 392–393)

Varela-Losada et al. (2016) pay special attention towards five aspects, which are: if the study is based on real participation by students; if the study promotes reflection on the complexity of environmental issues; if it facilitates critical thinking; if it encourages autonomous and responsible decision-making; and involves communities. These five aspects articulate a demand on practices of teaching and learning to reach out for and transform the participating teachers and students.

By presenting the four models by Eilks (2015), the 15 currents by Sauvé (2005), the three pillars by Giddings et al. (2002) and research discussing paradigmatic changes of teaching practices, we have illustrated a complex and divergent research field. In this book, in order to suit their respective aims, the authors of the chapters take points of departure in different parts of this literature. As pointed out, the aim of this chapter is to explore how environmental issues and sustainability have been/are conceptualised in relation to teacher education and the fields of science and technology, in South Africa and Sweden covering a period of 40 years, from 1980 to 2020. As part of the exploration, we sought to unravel the topics, models and theoretical approaches that have been discussed in research publications, focusing on the two countries. Our inquiry sought to tease out the main themes addressed, changes that can be identified over the years and differences and similarities in publications from North (Sweden) and South (South Africa) perspectives. The research questions guiding our exploration and analysis are:

1 How is science and technology teacher education addressed?
2 What theoretical approaches are identified?

## Method

In order to answer the questions paused above, we reviewed research publications in the broad field of education and sustainable development for the period 1980–2020 focusing research done in South Africa and Sweden. A data sample was constructed through a beta version of a search facility specialised to document the ESD field (Carlsson, Jarneving, Kullman, Omrcen, & Strömberg, 2020). The search facility was constructed using search phrases associated to 33 topic areas in order to identify all papers related to Sustainable Development. The search string EDUCATION was selected, replacing "reading" and "mathematics" with "science" and "technology" in the original search string, combined with teach* as a separate topic word. The new search string was applied to the database Web of Science. This search resulted in 1831 publications in total,

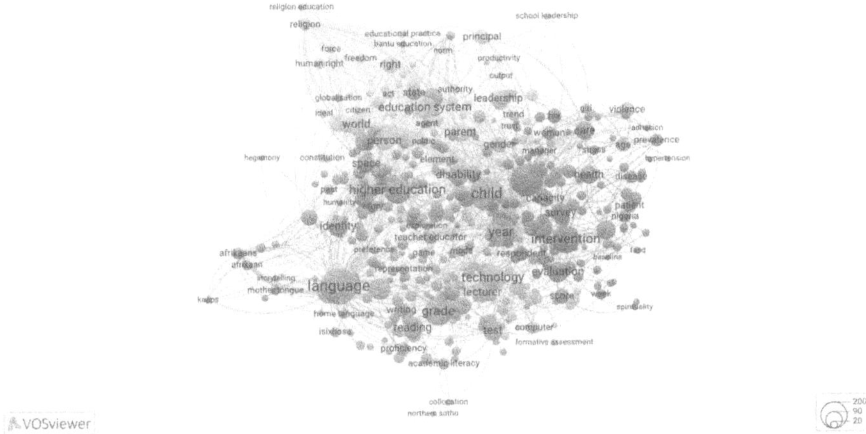


*Figure 1.2* Co-occurrence of frequent terms (494 terms occurring 12 times or more), in 1368 publications from South Africa between 1980 and 2019. Representation made in VOSviewer 1.6.15

1368 publications in South Africa and 463 publications in Sweden. The large sample of articles was imported to the software VOSviewer 1.6.15 (http://www.vosviewer.com) to produce representation of associated terms for the two countries (see Figures 1.2–1.4 and Table 1.1 in Appendix). In an exclusion process, the large sample was combined with a search for the word "teacher education", which resulted in a total result of 181 publications between 1980 and 2020. The titles and abstracts of the 181 publications were read through and those with a too general foci were found not relevant and excluded: teacher education in

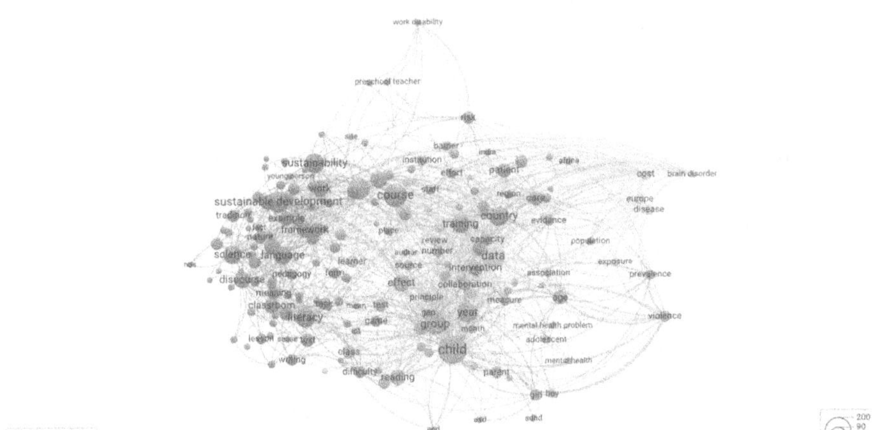


*Figure 1.3* Co-occurrence of frequent terms (193 terms occurring 12 times or more), in 463 publications from Sweden between 1980 and 2019. Representation made in VOSviewer 1.6.15

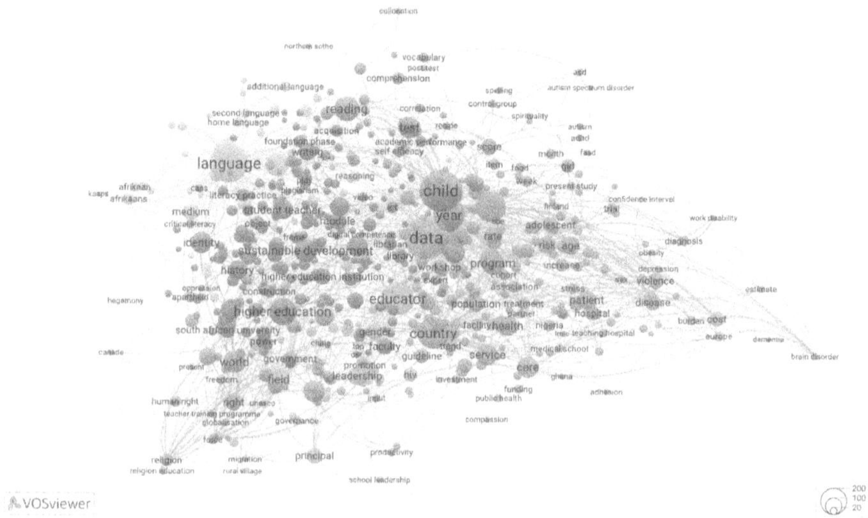


*Figure 1.4* Co-occurrence of frequent terms (615 terms occurring 12 times or more), in 1831 publications from South Africa and Sweden between 1980 and 2019. Representation made in VOSviewer 1.6.15

general (85), sustainability in higher education (6) and other education (34). The titles and abstracts of the relevant publications in science and technology teacher education (N = 44), and science, technology, engineering and mathematics (STEM) education (N = 12), were scrutinised one more time in relation to the purpose of the review; 27 were considered not relevant and 4 publications were not available online, leaving 25 publications selected as the primary sample. This sample was approached and analysed using content analysis. In summary, the criteria for inclusion in the sample were as follows:

- the article should focus on environmental, economic or ecologic issues in line with the SDGs (search facility Gothenburg University Library);
- focus on one of the two selected countries, South Africa and Sweden (selection made in the Web of Science database);
- focus on teacher education (evaluated from reading Abstract);
- focus on the education in the fields of science (biology, physics, chemistry) and technology (evaluated from reading Abstract);
- the articles should also be either empirical research studies or theoretical papers (use coding scheme).

The content analysis focused on how the article in a broad sense was addressing some of the 17 SDGs, the institution that was involved in the empirical papers, the targeted age level, if teachers or students were participants in a study, and what school subject or university discipline was in focus. Primarily, the content analysis tried to establish the theoretical approach, and if possible, to which of the 15 currents (Sauvé), or 16 including post-humanism, the article could be seen as associated to. We also considered if the international or global perspective was present and discussed. Finally, we examined how sustainability was conceptualised, and if models of implementing ESD in teacher education were mentioned (i.e. Eilks, 2015; Sauvé, 2005), as well as if examples of barriers and perceptions that prevent or support education to include ESD were discussed.

## Results

When the search string for the topic area EDUCATION was applied to all databases of Web of Science, it resulted in 1831 international publications between the years 1980 and 2020. The geographical distribution within those numbers was 1368 and 463 publications in South Africa and Sweden, respectively. Roughly, over each ten-year period, from 1980 to 2019, the total number of publications in South Africa is about three times the number of publications in Sweden, and in both countries the number of publications increases after the year 2000.

From the content analysis, the reviewed studies were categorised into six areas of foci which are: publications focusing on particular content; studies with interdisciplinary approaches; studies carrying with them implications for science and technology teacher education; publications calling for and discussing significant changes needed in teaching practices; publications focusing on challenging current paradigms in teaching and learning; and publications focusing on providing theoretical models on how to effectively integrate science and technology and sustainability in teacher education. The results are discussed in relation to what contributions may emerge from enhanced international collaborations between practitioners in the education of science and technology teachers.

### *Studies with content focus*

An example of studies sharing a content-focused approach towards education and sustainability is the one by Palmberg et al. (2015). The authors of this study claim that there is a strong relation between lack of understanding about species and lack of knowledge about sustainable development. Likewise, Palmberg et al. (2018) emphasise biodiversity education as an essential part of sustainability education that should be incorporated into teacher education programmes. It should be noted that this study marks a shift from biodiversity conservation to emotional and recreational arguments and aspects. These studies from Sweden show similar results to studies from elsewhere in the world, such as, for example, Park and Han (2016) that assesses the possibility of using robots in ESD by providing them to children from low-income families in South Korea. Another study from the USA by Lee, Nugent, Kunz, Houston and DeChenne-Peters (2018), emphasises the need to integrate inquiry and practice and incorporate engineering into science education with effective professional development as a critical component, which is a significant challenge for rural schools. In

Malaysia, Ling, Pang and Lajium (2019) report on the planning of after-school STEM education programme focusing on relevant global issues related to SDGs. A study by Wahono and Chang (2019) found that even if science teachers have a very good attitude towards ESD, they are moderate in their application and low in knowledge regarding STEM education.

### *Studies of interdisciplinary approaches*

Overall, studies sharing interdisciplinary approaches to science and technology teacher education suggest specific programmes or curricular units that involve opportunities for interdisciplinary topics or themes. Interdisciplinary approaches also include studies investigating students' classroom discussions about authentic problems so-called socioscientific issues (SSI), such as the preservation of the wolf (Åkerblom & Lindahl, 2017), or dilemmas regarding human sexuality and relationships within the topic of sex education (Lindahl & Lundin, 2016). Interdisciplinary approaches are also found in publications elsewhere in the world. For example, Suh and Han (2019) suggest an interdisciplinary approach to mathematical modelling which enables considerations of current and future real-world situations. In a related study, York, Lavi, Dori and Orgill (2019) suggest to include a systems thinking framework in chemistry education. A study by Stacchiotti, Acqua, Pennesi, Beccaceci and Paris (2019) investigates an activity about waste and circular economy. In this activity, topics like geo-resources, ore deposits and mining are used, to explain the life cycle of packaging materials.

### *Significant changes in teaching practices*

A number of studies point to the necessity of changing teaching practices significantly. One example is De Waal and Grösser (2009), who propose the importance of considering the learner's perspective, linking teaching and learning with students' learning styles. In their study, they found that teachers had difficulties in aligning their pedagogical approaches to accommodate learners' learning needs and fundamental rights. This was partly attributed to deficient teaching equipment. Other studies point towards the need to find strategies to empower educators (November, Alexander, & Van Wyk, 2010) or enhance capacity and agency of teacher educators, for example, Mandikonza and Lotz-Sisitka (2016). In similar vein, Robinson and McMillan (2006) emphasise cultural and social aspects, claiming the role of teacher educators themselves as being one barrier for change, and time another.

### *Studies challenging paradigms*

One group of studies challenges current paradigms in teaching and learning, envisioning an alteration in perspectives on knowledge and education. For example, Ogunniyi (2007) articulates that in order to overcome the negative impact of our dominant system, a joint study of western knowledge systems and indigenous knowledge systems is a necessity. With a similar interest Mudaly (2018) and Adebayo and Mudaly (2019) argue for decolonising the science teacher education curriculum. Another study by Ramnarain and Moleki (2017) suggests the implementation of a humanistic perspective on science using newspapers, claiming against a transmission-model, that professional development programmes demand for renewed conception of scientific literacy.

### *Theoretical models*

Two studies present theoretical models. Keinonen and de Jager (2017) present the model for culturally responsive teaching. The model defines five thematic categories: content integration, facilitating knowledge construction, prejudice reduction, social justice and academic development. In a similar study, Kibuka-Sebitosi (2015) presents the collaboration model. This model is conceptualised along four pillars: (i) the international knowledge base (Mexico, India, German and South Africa); (ii) the local partnerships (NGOs, universities, private sector, government); (iii) the shared values, attitudes and practices on sustainability; and (iv) the underlying principle of the whole school approach, that is multi-, inter- and transdisciplinary in nature.

### *Perspectives on effectiveness*

One group of studies provides evaluations of effectiveness of integrating science and technology and sustainability in curricula. For example, Treu, Olivier, Bean and Van der Walt (2010) investigate an implementation process of new curricula through a framework of ten dynamics. The study finds difficulties and three themes confirming a gap between national and provincial levels, and a fourth theme where teachers show readiness to undertake implementation despite problems. Loubser (2015) argues for EE and education for sustainability in an interdisciplinary way in student teachers. She suggests that the skills and knowledge student teachers gain should be assessed during teaching practice visits by lecturers. Andersson, Jagers, Lindskog and Martinsson (2013) and Andersson (2017) investigate ESD in teacher education and whether effects on student teachers' perceptions of how to teach pupils about sustainable development can be found. Both studies find short-term effects towards student teachers actively getting pupils to discuss issues concerning the environment but without influencing their opinions about it.

## Discussion

As we noted, in this chapter, our major aim was to investigate how science and technology teacher education was addressed in South Africa and Sweden, gleaning from the research literature. We also sought to identify the theoretical approaches guiding research in science and technology teacher ESD. Within the ambit of these two broad goals of our objective, we also sought to provide North and South perspectives on how issues of sustainability have been conceptualised in relation to the education of teachers in natural sciences and technology, in particular, and the education of natural sciences and technology, in general. As part of this endeavour, we were able to unstitch the main themes addressed as well as the changes, differences and similarities in ESD as revealed by research publications from both countries.

### *ESD and science and technology teacher education in South Africa and Sweden*

Our bibliographic analysis shows that, in the two countries, an immense number of publications addressing ESD have been produced, especially during the past ten years. Our content analysis revealed clusters of main themes and topics addressed. Further, our content analysis reveals that for both South Africa and Sweden, research on ESD can be categorised into six areas of foci, which are; studies focusing on content, studies of interdisciplinary approaches, studies focusing on significant changes in teaching practices, studies challenging paradigms, studies focusing on theoretical models, and perspectives and evaluations of teaching and learning effectiveness. Overall, the results indicate that these issues are of concern to researchers both in the North and the South. We believe that this is so because both South Africa and Sweden are facing similar sustainability challenges. In both countries, some researchers are preoccupied with the major issues and challenges of how best teacher education can contribute to sustainability and achievement of UN SDGs. However, only a few articles explicitly mention issues of sustainability in relation to the education of science and technology teachers. Moreover, very few studies explicitly focus on how the science and/or technology teachers' practices can be developed in order to promote their implementation of ESD in the classroom. This is of major concern given that it is generally accepted that teachers are expected to have many complex competencies (Wiek, Withycombe, & Redman, 2011), including pedagogical content knowledge (Brandt, Bürgener, Barth, & Redman, 2019), if they are to effectively participate in teaching sustainable development and contributing towards achievement of UN SDGs.

While the research on ESD in South Africa and Sweden has broadly fitted into the six areas of foci identified above, a closer analysis of the 25 publications (19 South Africa and 6 Sweden) from both counties reveals some nuances of difference in terms of what researchers in each country have concentrated on. Our argument is that these observed distinctions can be taken as indicators of a difference in the way in which ESD is approached in each of the two countries. In Sweden, a large part of the published research has focused on understanding and evaluating teachers' and school learners' interest and competencies (knowledge, attitudes and behaviours) in sustainability (see, for example, Besong & Holland, 2015; Frisk & Larson, 2011). On the other hand, researchers in South Africa have focused largely on understanding pedagogical practices in teaching and learning ESD, analysis of policy and integration of ESD into mainstream curricula. For South Africa, this leaning in focus could have been driven by the challenges faced in adapting teaching and curricula to global trends and driven by the desire for decolonisation of education in the post-apartheid era. We argue that these observed nuances of distinction are instructive of the differences in approaches and strategies to tackling sustainability challenges between the North (Sweden) and the South (South Africa). For Sweden, the emphasis is on understanding the levels of ESD competencies among teachers, student teachers and school learners, as a basis for developing their scientific technological literacy and consequently enhancing their contributions to sustainability. As Fredriksson, Kusanagi, Gougoulakis, Matsuda and Kitamura (2020) posit, in Sweden, developing sustainability is more a matter of mentioning ESD as part of the principles that guide education. In contrast, in South Africa, the emphasis is on understanding and/or developing teachers' pedagogical competences in implementing policies and curricula, including integration of ESD.

### *Theoretical perspectives to ESD in South Africa and Sweden*

Our analysis reveals that the research in both countries has been underpinned by a wide range of theoretical perspectives such as cognitive perspectives (Pavlova, 2013) and critical theory and socio-ecological theory (see, for example, Aguayo & Eames, 2017; Lockhart, 2018) and in many instances spiced with ideas about globality (Ahamer & Sayamov, 2013). Further, this analysis shows that in both countries, researchers have almost equally given attention to both currents with a longer tradition in EE and currents that more recently emerged (Sauvé, 2005). Profoundly, in the two countries researchers have advocated for both changes in individual behaviour or collective action and the importance of human sustainable use of the environment and the earth's resources. Overall, theoretically, the majority of the analysed publications recognise the existence as well as the interconnectedness among the three pillars of sustainability – the economic, social and environmental. It appears that in both the North and the South, the "three pillars approach" to sustainability is the dominant paradigm.

While there are these evident commonalities, there are also some subtleties peg marking recognisable differentials in the theoretical underpinnings in ESD research in South Africa and Sweden. As alluded to, in South Africa, both research and implementation of ESD in teacher education have responded to calls for indigenisation and decolonisation of education (see, for example, Mokuku, 2012; Shallcross & Robinson, 2007). There are arguments that both the content and pedagogy of ESD in teacher education must respond to the socioeconomic, political and ideological realities of the post-apartheid era. Issues of relevance and colonial hegemony in education permeate this debate. On the other hand, in Sweden, the discourse on ESD provision in teacher education is increasingly under the influence if not domination of systems thinking theories and perspectives (see, for example. Palmberg, Hofman-Bergholm, Jeronen, & Yli-Panula, 2017; Fredriksson et al., 2020). It is argued that in order to understand the complexity of sustainability, it is necessary to focus on and understand how the interrelated pillars of sustainability (social, economic, environmental, etc.) holistically shape and contribute to achievement of sustainable development. Of course, there are also other several perspectives and theories agitating for legitimacy in this discourse. Post-humanism (see, for example, Weldemariam, 2017), anthropocentric and non-anthropocentric sub-positions (Kronlid & Öhman, 2013) are cases in point.

## Conclusions and recommendations

Our analysis of the research literature on EE and ESD focusing on science and technology teacher education in South Africa and Sweden has revealed that the countries face similar challenges in sustainability. It is noteworthy that both countries show commitments to achieving sustainable development through implementation of the UN SDGs by developing teachers' scientific and technological literacy. Evidently, researchers in science and technology education in the two countries have generally been involved in similar areas of research focus. While in both countries the research efforts have been underpinned by a wide range of dominant theoretical paradigms including cognitive theories, critical theory and socio-ecological theory, the research appears to be driven by different ideological and philosophical orientations and beliefs. Whereas in the North (Sweden), the research is attuned to neo-liberal and Western imperatives, in South (South Africa), the bulk of the research appears to be a response to socio-political voices calling for decolonisation of education and the indigenisation of policies (including IKS) and curricula so as to make ESD germane to the realities of the post-colonial era. Albeit, the countries are facing very similar, environmental and sustainability challenges.

Although the two countries agree that contemporary challenges emanating from unsustainable exploitation of the environment and the earth's resources necessitate reorientation in science and technology teacher education, there are differences in the ways in which each country approaches ESD. In Sweden, the emphasis has been on understanding and developing teachers' and learners' sustainability competencies, while in South Africa the major focus has been on developing teachers' pedagogical practices and implementation of policies and curricula.

Given the insights we have gleaned from this comparative study, our contention is that, in this era of the Anthropocene, countries in both the North and the South can benefit the cause sustainability through sharing experiences, comparing practices and discussing possibilities and solutions to common problems. Going into the future, we recommend studies focusing on comparing levels of teachers' and students' sustainability competencies. Such studies will help answer such an interesting question as: How big is the sustainability competencies gap between teachers/students in the North and the South? A large-scale international comparative study of sustainability competencies will not only generate global interest in sustainability in science and technology education, but also awaken politicians, policymakers and teacher educators to the seriousness of sustainable development and slowing humans' galloping of the earth towards cataclysm.

## Appendix

*Table 1.1* Sample of 25 Publications

| | *Reference* | *Country, year* |
|---|---|---|
| 1 | Modiba, M. (1997). Distance teacher education in South Africa: A critical analysis of pedagogical assumptions. *Teaching and Teacher Education, 13*(7), 727–739. | *South Africa, 1997* |
| 2 | Robinson, M., & McMillan, W. (2006). Who teaches the teachers? Identity, discourse and policy in teacher education. *Teaching and Teacher Education, 22*(3), 327–336. | *South Africa, 2006* |
| 3 | De Waal, E., & Grösser, M. (2009). Safety and security at school: A pedagogical perspective. *Teaching and Teacher Education, 25*(5), 697–706. | *South Africa, 2009* |
| 4 | Alexander, G., Van Wyk, M. M., Bereng, T., & November, I. (2010). Legitimate Peripheral Participation (LPP)–The case for Recognition of Prior Learning sites and knowledges in South Africa's transforming education system. *Teaching and Teacher Education, 26*(1), 45–52. | *South Africa, 2010* |
| 5 | November, I., Alexander, G., & Van Wyk, M. M. (2010). Do principal-educators have the ability to transform schools?: A South African perspective. *Teaching and Teacher Education, 26*(4), 786–795. | *South Africa, 2010* |
| 6 | Treu, P., Olivier, M., Bean, P., & Van der Walt, J. (2010). A group of educators' stance on the implementation of South Africa's further education and training curriculum. *Teaching and Teacher Education, 26*(2), 343–350. | *South Africa, 2010* |
| 7 | Ogunniyi, M. B., & Rollnick, M. (2015). Pre-service science teacher education in Africa: Prospects and challenges. *Journal of Science Teacher Education, 26*(1), 65–79. doi:10.1007/s10972-014-9415-y | *South Africa, 2015* |
| 8 | Palmberg, I., Berg, I., Jeronen, E., Kärkkäinen, S., Norrgård-Sillanpää, P., Persson, C.,. . . Yli-Panula, E. (2015). Nordic–Baltic student teachers' identification of and interest in plant and animal species: The importance of species identification and biodiversity for sustainable development. *Journal of Science Teacher Education, 26*(6), 549–571. doi:10.1007/s10972-015-9438-z | *Sweden, 2015* |
| 9 | Lindahl, M. G., & Lundin, M. (2016). How do 15–16 year old students use scientific knowledge to justify their reasoning about human sexuality and relationships? *Teaching and Teacher Education, 60*, 121–130. doi:https://doi.org/10.1016/j.tate.2016.08.009 | *Sweden, 2016* |
| 10 | Åkerblom, D., & Lindahl, M. (2017). Authenticity and the relevance of discourse and figured worlds in secondary students' discussions of socioscientific issues. *Teaching and Teacher Education, 65*, 205–214. | *Sweden, 2016* |
| 11 | Keinonen, T., & de Jager, T. (2017). Student teachers' perspectives on chemistry education in South Africa and Finland. *Journal of Science Teacher Education, 28*(6), 485–506. | *South Africa (Finland), 2017* |
| 12 | Ramnarain, U., & Moleki, B. (2017). Teachers' use of newspaper articles in promoting a humanistic perspective of science in South Africa. *Journal of Science Teacher Education, 28*(2), 205–217. | *South Africa, 2017* |
| 13 | Andersson, K., Jagers, S. C., Lindskog, A., & Martinsson, J. (2013). Learning for the future? Effects of education for sustainable development (ESD) on teacher education students. *Sustainability, 5*(12), 5135–5152. doi:10.3390/su5125135 | *Sweden, 2013* |

(*Continued*)

*Table 1.1* Sample of 25 Publications (*Continued*)

| | *Reference* | *Country, year* |
|---|---|---|
| 14 | Kibuka-Sebitosi, E. (2015). Towards a model for international collaboration and partnerships in teacher education in Africa: Education for sustainable development in South Africa. *International Journal of African Renaissance Studies, 10*(2), 104–124. doi:10.1080/18186874.2015.1107983 | *South Africa, 2015* |
| 15 | Loubser, C. P. (2015). Students' views about the inclusion of environmental education and education for sustainability in teacher education courses. *International Journal of Educational Sciences, 8*(1), 93–100. | *South Africa, 2015* |
| 16 | Peden, M. (2015). Education for sustainable agriculture: The old and new curricula for agricultural sciences in South African schools. *Journal of Education,* (60), 63–99. | *South Africa, 2015* |
| 17 | Ramnarain, U. (2015). A Pilot Study on the Mentoring of PGCE Physical Sciences Students at a Teaching School. In Y. Suan & L. U. Hua (Eds.), *Xvi International Organisation for Science and Technology Education Symposium* (Vol. 167, pp. 44–49). | *South Africa, 2015* |
| 18 | Mandikonza, C., & Lotz-Sisitka, H. (2016). Emergence of environment and sustainability education (ESE) in teacher education contexts in Southern Africa: A common good concern. *Educational Research for Social Change, 5*(1), 107–130. | *South Africa, 2016* |
| 19 | Andersson, K. (2017). Starting the pluralistic tradition of teaching? Effects of education for sustainable development (ESD) on pre-service teachers' views on teaching about sustainable development. *Environmental Education Research, 23*(3), 436–449. doi:10.1080/13504622.2016.1174982 | *Sweden, 2017* |
| 20 | Mudaly, R. (2018). Towards decolonising a module in the pre-service science teacher education curriculum: The role of indigenous knowledge systems in creating spaces for transforming the curriculum. *Journal of Education,* (74), 47–66. doi:10.17159/2520-9868/i74a04 | *South Africa, 2018* |
| 21 | Palmberg, I., Hermans, M., Jeronen, E., Karkkainen, S., Persson, C., & Yli-Panula, E. (2018). Nordic student teachers' views on the importance of species and species identification. *Journal of Science Teacher Education, 29*(5), 397–419. doi:10.1080/1046560x.2018.1468167 | *Sweden, 2018* |
| 22 | Adebayo, O. A., & Mudaly, R. (2019). Creating a decolonised curriculum to address food insecurity among university students. *Problems of Education in the 21st Century, 77*(2), 179–194. doi:10.33225/pec/19.77.179 | *South Africa, 2019* |
| 23 | Uleanya, C., Uleanya, M. O., & Olayemi, S. (2019). Revisiting high school teachers' education: Solution to sustainable development in rural Africa. *Cogent Social Sciences, 5*(1). doi:10.1080/23311886.2019.16535 45 | *South Africa, 2019* |
| 24 | Ogunniyi, M. B. (2007). Teachers' stances and practical arguments regarding a science-indigenous knowledge curriculum: Part 2. *International Journal of Science Education, 29*(10), 1189–1207. doi:10.1080/09500690600931038 | *South Africa, 2007* |
| 25 | Onwu, G., & Stoffels, N. (2005). Instructional functions in large, under-resourced science classes: Perspectives of South African teachers. *Perspectives in Education, 23*(3), 79–91. | *South Africa, 2005* |

## References

Adebayo, O. A., & Mudaly, R. (2019). Creating a decolonised curriculum to address food insecurity among university students. *Problems of Education in the 21st Century*, *77*(2), 179.

Aguayo, C., & Eames, C. (2017). Promoting community socio-ecological sustainability through technology: A case study from Chile. *International Review of Education*, *63*(6), 871–895.

Ahamer, G., & Sayamov, Y. N. (2013). Education as a global "soft power" for sustainable development. *Campus-Wide Information Systems*, *30*(5), 346–357.

Åkerblom, D., & Lindahl, M. (2017). Authenticity and the relevance of discourse and figured worlds in secondary students' discussions of socioscientific issues. *Teaching and Teacher Education*, *65*, 205–214.

Alaimo, S. (2012). Sustainable this, sustainable that: New materialisms, posthumanism, and unknown features. (Theories and methodologies) (Essay). *PMLA*, *127*(3), 558–564.

Andersson, K. (2017). Starting the pluralistic tradition of teaching? Effects of education for sustainable development (ESD) on pre-service teachers' views on teaching about sustainable development. *Environmental Education Research*, *23*(3), 436–449.

Andersson, K., Jagers, S. C., Lindskog, A., & Martinsson, J. (2013). Learning for the future? Effects of education for sustainable development (ESD) on teacher education students. *Sustainability*, *5*(12), 5135–5152.

Barratt Hacking, E., Barratt, R., & Scott, W. (2007). Engaging children: Research issues around participation and environmental learning. *Environmental Education Research*, *13*(4), 529–544.

Berglund, T., & Gericke, N. (2016). Separated and integrated perspectives on environmental, economic, and social dimensions – an investigation of student views on sustainable development. *Environmental Education Research*, *22*(8), 1115–1138.

Besong, F., & Holland, C. (2015). The dispositions, abilities and behaviours (DAB) framework for profiling learners' sustainability competencies in higher education. *Journal of Teacher Education for Sustainability*, *17*(1), 5–22.

Brandt, J. O., Bürgener, L., Barth, M., & Redman, A. (2019). Becoming a competent teacher in education for sustainable development. *International Journal of Sustainability in Higher Education*.

Carlsson, H., Jarneving, B., Kullman, L., Omrcen, E., & Strömberg, D. (2020). *Gothenburg Sustainable Development Search Strings 2019*. Retrieved from http://hdl.handle.net/2077/64070

Carson, R. (2002). *Silent spring* (40th anniversary ed.). Boston, MA: Mariner Books.

Colantonio, A. (2009) Social sustainability: A review and critique of traditional versus emerging themes and assessment methods. In Horner, M., Price, A., Bebbington, J., & Emmanuel, R. (Eds.), *Sue-Mot Conference 2009: Second International Conference on Whole Life Urban Sustainability and Its Assessment: Conference Proceedings*. Loughborough University, Loughborough, pp. 865–885. ISBN 9780947974.

Dahl Madsen, K. (2013). Unfolding education for sustainable development as didactic thinking and practice. *Sustainability*, *5*(9), 3771–3782. doi: 10.3390/su5093771

De Waal, E., & Grösser, M. (2009). Safety and security at school: A pedagogical perspective. *Teaching and Teacher Education*, *25*(5), 697–706.

Eilks, I. (2015). Science education and education for sustainable development – justifications, models, practices and perspectives. *Eurasia Journal of Mathematics, Science and Technology Education*, *11*(1), 149–158.

Fuertes-Camacho, M. T., Graell-Martín, M., Fuentes-Loss, M., & Balaguer-Fàbregas, M. C. (2019). Integrating sustainability into higher education curricula through the project method, a global learning strategy. *Sustainability, 11*(3), 767.

Fredriksson, U., Kusanagi, K. N., Gougoulakis, P., Matsuda, Y., & Kitamura, Y. (2020). A comparative study of curriculums for education for sustainable development (ESD) in Sweden and Japan. *Sustainability, 12*(3), 1123.

Frisk, E., & Larson, K. L. (2011). Educating for sustainability: Competencies & practices for transformative action. *Journal of Sustainability Education, 2*(1), 1–20.

Giddings, B., Hopwood, B., & O'Brien, G. (2002). Environment, economy and society: Fitting them together into sustainable development. *Sustainable Development, 10*, 187–196. doi: 10.1002/sd.199

Hamilton, C. (2017). *Defiant earth: The fate of humans in the Anthropocene*. Oxford: John Wiley & Sons.

Jensen, B. B., & Schnack, K. (1997). The action competence approach in environmental education. *Environmental Education Research, 3*(2), 163–178.

Keinonen, T., & de Jager, T. (2017). Student teachers' perspectives on chemistry education in South Africa and Finland. *Journal of Science Teacher Education, 28*(6), 485–506.

Kibuka-Sebitosi, E. (2015). Towards a model for international collaboration and partnerships in teacher education in Africa: Education for sustainable development in South Africa. *International Journal of African Renaissance Studies – Multi-, Inter- and Transdisciplinarity, 10*(2), 104–124.

Knutsson, B. (2013). Swedish environmental and sustainability education research in the era of post-politics? *Utbildning & Demokrati–tidskrift för didaktik och utbildningspolitk, 22*(2), 105–122.

Kronlid, D. O., & Öhman, J. (2013). An environmental ethical conceptual framework for research on sustainability and environmental education. *Environmental Education Research, 19*(1), 21–44.

Kyburz-Graber, R. (2013). 3 Socioecological Approaches to Environmental Education and Research: A Paradigmatic Response to Behavioral Change Orientations. In R. B. Stevenson, M. Brody, J. Dillon & A. E. J. Wals (Eds) *International handbook of research on environmental education* (pp. 23–32). New York: Routledge. https://doi.org/10.4324/9780203813331

Lave, J., and E. Wenger. 1991. *Situated learning: Legitimate peripheral participation*. Cambridge: Cambridge University Press. doi: 10.1017/CBO9780511815355

Lee, S. C., Nugent, G., Kunz, G. M., Houston, J., & DeChenne-Peters, S. (2018). Case study: Value-added benefit of distance-based instructional coaching on science teachers' inquiry instruction in rural schools. *Journal of Science Teacher Education, 29*(3), 179–199.

Lehtonen, M. (2004). The environmental–social interface of sustainable development: Capabilities, social capital, institutions. *Ecological Economics, 49*(2), 199–214.

Lindahl, M. G., & Lundin, M. (2016). How do 15–16 year old students use scientific knowledge to justify their reasoning about human sexuality and relationships? *Teaching and Teacher Education, 60*, 121–130. doi: 10.1016/j.tate.2016.08.009

Ling, L. S., Pang, V., & Lajium, D. (2019). The planning of integrated STEM education based on standards and contextual issues of sustainable development goals (SDG). *Journal of Nusantara Studies (JONUS), 4*(1), 300–315.

Lockhart, A. S. (2018). Monitoring ESD: Lessons learned and ways forward. In A. Leicht, J. Heiss & W. J. Byun (Eds), *Issues and Trends in Education for Sustainable Development* (215–231). Paris: UNESCO Publishing.

Loubser, C. P. (2015). Students' views about the inclusion of environmental education and education for sustainability in teacher education courses. *International Journal of Educational Sciences, 8*(1), 93–100.

Mandikonza, C., & Lotz-Sisitka, H. (2016). Emergence of environment and sustainability education (ESE) in teacher education contexts in Southern Africa: A common good concern. *Educational Research for Social Change*, *5*(1), 107–130.

McCallum, B., Hargreaves, E., & Gipps, C. (2000). Learning: The pupil's voice. *Cambridge Journal of Education*, *30*(2), 275–289.

Mogensen, F. & Mayer, M. (2005). *Eco-schools: Trends and divergences. A comparative study on ECO-school development processes in 13 countries.* Viena: Austrian Federal Ministry of Education, Science and Culture. http://www.ubu10.dk/downloadfiles/comparative2.pdf

Mogensen, F., & Schnack, K. (2010). The action competence approach and the 'new' discourses of education for sustainable development: Competence and quality criteria. *Environmental Education Research*, *16*(1), 59–74.

Mokuku, T. (2012). Lehae La Rona: Epistemological interrogation to broaden our conception of environment and sustainability. *Canadian Journal of Environmental Education (CJEE)*, *17*, 159–172.

Mudaly, R. (2018). Towards decolonising a module in the pre-service science teacher education curriculum: The role of indigenous knowledge systems in creating spaces for transforming the curriculum. *Journal of Education (University of KwaZulu-Natal)*, (74), 47–66.

November, I., Alexander, G., & Van Wyk, M. M. (2010). Do principal-educators have the ability to transform schools?: A South African perspective. *Teaching and Teacher Education*, *26*(4), 786–795.

Ogunniyi, M. (2007). Teachers' stances and practical arguments regarding a science-indigenous knowledge curriculum: Part 2. *International Journal of Science Education*, *29*(10), 1189–1207.

Palmberg, I., Berg, I., Jeronen, E., Kärkkäinen, S., Norrgård-Sillanpää, P., Persson, C. … Yli-Panula, E. (2015). Nordic–Baltic student teachers' identification of and interest in plant and animal species: The importance of species identification and biodiversity for sustainable development. *Journal of Science Teacher Education*, *26*(6), 549–571. doi: 10.1007/s10972-015-9438-z

Palmberg, I., Hermans, M., Jeronen, E., Kärkkäinen, S., Persson, C., & Yli-Panula, E. (2018). Nordic student teachers' views on the importance of species and species identification. *Journal of Science Teacher Education*, *29*(5), 397–419.

Palmberg, I., Hofman-Bergholm, M., Jeronen, E., & Yli-Panula, E. (2017). Systems thinking for understanding sustainability? Nordic student teachers' views on the relationship between species identification, biodiversity and sustainable development. *Education Sciences*, *7*(3), 72.

Park, I.-W., & Han, J. (2016). Teachers' views on the use of robots and cloud services in education for sustainable development. *Cluster Computing*, *19*(2), 987–999.

Pavlova, M. (2013). Teaching and learning for sustainable development: ESD research in technology education. *International Journal of Technology and Design Education*, *23*(3), 733–748.

Purvis, B., Mao, Y., & Robinson, D. (2019). Three pillars of sustainability: In search of conceptual origins. *Sustainability Science*, *14*(3), 681–695.

Ramnarain, U., & Moleki, B. (2017). Teachers' use of newspaper articles in promoting a humanistic perspective of science in South Africa. *Journal of Science Teacher Education*, *28*(2), 205–217.

Robinson, M., & McMillan, W. (2006). Who teaches the teachers? Identity, discourse and policy in teacher education. *Teaching and Teacher Education*, *22*(3), 327–336.

Sandell, K., & Öhman, J. (2010). Educational potentials of encounters with nature: Reflections from a Swedish outdoor perspective. *Environmental Education Research*, *16*(1), 113–132. doi: 10.1080/13504620903504065

Sauvé, L. (2005). Currents in environmental education: Mapping a complex and evolving pedagogical field. *Canadian Journal of Environmental Education, 10*(1), 11–37.

Sepkoski, D. (2020). *Catastrophic thinking: Extinction and the value of diversity from Darwin to the Anthropocene.* Chicago, IL: University of Chicago Press.

Shallcross, T., & Robinson, J. (2007). Is a decade of teacher education for sustainable development essential for survival? *Journal of Education for Teaching, 33*(2), 137–147.

Stacchiotti, L., Acqua, A., Pennesi, D., Beccaceci, A., & Paris, E. (2019). Wasteberg: A didactic activity about waste and sustainable use of georesources in relation to the Agenda 2030. *Rendiconti Online della Società Geologica Italiana, 49*, 127–133. doi:10.3301/ROL.2019.62

Stevenson, R. B. & Stirling, C. (2010). Environmental learning and agency in diverse educational and cultural contexts. In R.B. Stevenson and J. Dillon (Eds), *Engaging environmental education: Learning, culture and agency*, (219–237). Rotterdam: Sense.

Suh, H., & Han, S. (2019). Promoting sustainability in university classrooms using a STEM project with mathematical modeling. *Sustainability, 11*(11), 3080.

Treu, P., Olivier, M., Bean, P., & Van der Walt, J. (2010). A group of educators' stance on the implementation of South Africa's further education and training curriculum. *Teaching and Teacher Education, 26*(2), 343–350.

Varela-Losada, M., Vega-Marcote, P., Pérez-Rodríguez, U., & Álvarez-Lires, M. (2016). Going to action? A literature review on educational proposals in formal environmental education. *Environmental Education Research, 22*(3), 390–421. doi: 10.1080/13504622.2015.1101751

Wahono, B., & Chang, C.-Y. (2019). Assessing teacher's attitude, knowledge, and application (AKA) on STEM: An effort to foster the sustainable development of STEM education. *Sustainability, 11*(4), 950.

Wals, A. (Ed). (2007). *Social Learning towards a Sustainable World.* Wageningen: Wageningen Academic Publishers.

Wals, A. E. J., Weakland, J., & Corcoran, P. B. (2017). Preparing for the ecocene: Envisioning futures for environmental and sustainability education. *Japanese Journal of Environmental Education, 26*(4), 71–76. doi: 10.5647/jsoee.26.4_71

Weldemariam, K. (2017). Challenging and expanding the notion of sustainability within early childhood education: Perspectives from post-humanism and/or new materialism. In *Ethical literacies and education for sustainable development* (pp. 10–126). Cham: Palgrave Macmillan.

Wiek, A., Withycombe, L., & Redman, C. (2011). Key competencies in sustainability: A reference framework for academic program development. *Sustainability Science, 6*(2), 203–218.

World Commission on Environment and Development (1987). Our common future. Nairobi: United Nations Environment Programme.

York, S., Lavi, R., Dori, Y. J., & Orgill, M. (2019). Applications of systems thinking in STEM education. *Journal of Chemical Education, 96*(12), 2742–2751.

Zalasiewicz, J., Waters, C. N., Summerhayes, C. P., Wolfe, A. P., Barnosky, A. D., Cearreta, A., & Williams, M. (2017). The Working Group on the Anthropocene: Summary of evidence and interim recommendations. *Anthropocene, 19*, 55–60. doi: 10.1016/j.ancene.2017.09.001

Zhang, Y., & Wang, P. (2021). Detecting the historical roots of education for sustainable development (ESD): A bibliometric analysis. Emerald Insight. https://www.emerald.com/insight/1467-6370.htm

# 2 Science and technology teacher education in South Africa and Sweden

## A comparison of linked policies and curricula

*Suresh Kamar Singh, Anna Maria Hipkiss, Portia Kavai, Kassahun Weldemariam, Nina Christenson and Washington Dudu*

### Introduction and background

Climate change, poverty, water scarcity, increasing global population, new diseases (e.g. COVID-19) and limited natural resources are just some of the challenges that nations face in the 21st century. These challenges have necessitated the question: *how do we ensure a sustainable future for all?* Inevitably, this has resulted in calls for transformative education such as education for sustainable development (ESD). In 2002, the United Nations General Assembly (UN) recognised ESD as being transformative when it passed Resolution 57/254 entitled, "United Nations Decade of Education for Sustainable Development (2005–2014)". The UN mandated the United Nations Educational, Scientific and Cultural Organization (UNESCO) to develop a plan that foregrounds education as the key driver in the promotion of sustainability as it identified education as fundamental to achieving sustainable development goals (SDGs). Sustainable development (SD), as described in the Brundtland Report, relates to creating harmony between three important spheres: natural resources, economics and social resources (Keeble, 1988). The UN initiative, a decade (2005–2014) of ESD, has passed. In 2014, UNESCO launched the Global Action Programme (GAP) on ESD in all areas of education. In 2015, the UN accepted the 2030 Agenda for sustainable development and adopted a set of SDGs with sustainable development goal 4 (SDG4), Education, identified as a stand-alone goal.

Post-2014 issues of sustainability and the Anthropocene (links to natural resources) persist. The UN's Intergovernmental Panel on Climate Change (IPCC) published a report on climate change for the 26th UN Climate Change Conference of the parties (COP26) in Glasgow 2021. The IPCC Report (2021) paints a bleak picture of future global temperatures showing human activity damaging the world at a rapid rate that will lead to more drought, heatwaves,

DOI: 10.4324/9781003190158-4

cyclones, typhoons and tsunamis. The report is described as a code red for humanity, whilst others have labelled it as terrifying. The report includes strong evidence that human activity has been the primary cause of global warming. Pollution, urbanisation and deforestation have contributed to the current climate crisis. Greta Thunberg, a climate change activist's impassioned plea to world leaders at the UN in September 2019 shows that the world has not made sufficient progress in addressing the major issues in sustainable development (SD). The key enabler for transforming society into more sustainable futures is ESD. Countries across the world have made significant strides in ESD. However, to avert the crises, there is a need for education to champion awareness on issues related to sustainability.

The UN points out that investment in ESD is an investment in the future (UNESCO, 2009b). Furthermore, the UN highlights:

> ESD requires an emphasis on new learning processes and educational methodologies as opposed to merely addressing SD topics and focusing on the transfer of knowledge about these topics. Schools, curriculum development institutes, and educational research organizations should be at the forefront of the search for and development of these new forms of learning and the kinds of curricula, learning environments, and school-community relationships that are needed to enable such learning to flourish...
>
> (UNESCO, 2009b, p. 65)

In its Bonn declaration after the UNESCO World Conference on ESD, the UN called for:

- Support the incorporation of sustainable development issues using an integrated and systemic approach in formal education as well as in non-formal and informal education at all levels, in particular through the development of effective pedagogical approaches, teacher education, teaching practice, curricula, learning materials, and education leadership development, and also by recognizing the significant contribution of non-formal education and informal learning as well as vocational and work-place learning. Sustainable development is a cross-cutting theme with relevance to all disciplines and sectors.
- Reorient curriculum and teacher education programmes to integrate ESD into both pre-service and in-service programmes. Support teacher education institutions, teachers, and professors to network, develop, and research sound pedagogical practices. Specifically support teachers to develop ESD strategies.

(UNESCO, 2009a)

Weldemariam et al. (2017, p. 335) acknowledge that although there is no agreed definition on what sustainability is, it can be described broadly as "a discipline that requires major efforts to ensure the well-being of people and the planet now and in the future". The initiative by the UN requires countries to integrate all aspects of SD in their teaching and learning. This implies that curricula reforms are inevitable, as all countries have to re-examine their schools and higher education curricula. The focus of this chapter is to analyse policy and curricula documents, so as to illuminate and elucidate the status quo, regarding ESD and science and technology education. In this regard, the approaches to ESD in teacher education policy and linked curricula in schools and at four conveniently sampled universities in South Africa and Sweden (two from each country) are examined. It is interesting to note that the two countries have different SGD Index scores and are at different ends of the SDG Index Table. South Africa has an SDG Index Score of 63.7 whilst Sweden has a score of 85.6 (Sachs, Kroll, Lafortune, Fuller, & Woelm, 2021, pp. 10–11).

## Education for sustainable development (ESD)

As alluded to, ESD is a UNESCO initiative aimed at transforming education for sustainability. According to the UNESCO:

> Education for Sustainable Development (ESD) empowers learners of all ages with the knowledge, skills, values, and attitudes to address the interconnected global challenges we are facing, including climate change, environmental degradation, loss of biodiversity, poverty, and inequality. Learning must prepare students and learners of all ages to find solutions for the challenges of today and the future. Education should be transformative and allow us to make informed decisions and take individual and collective action to change our societies and care for the planet.
>
> (UNESCO, 2013, p. 1)

The definition of ESD adopted in this chapter is as follows:

> ...ESD is a broad concept bringing a distinctive orientation to many important aspects of education on the whole, including access, relevance, equity, and inclusivity. Thus, ESD is far more than teaching knowledge and principles related to sustainability. ESD, in its broadest sense, is education for social transformation to create more sustainable societies. ESD touches every aspect of education including planning, policy development, programme implementation, finance, curricula, teaching, learning, assessment, administration. ESD aims to provide a coherent interaction between education, public awareness, and training to create a more sustainable future...
>
> (UNESCO, 2012, p. 33)

This aligns with SDG (Goal 4.7) which states that:

> ...all learners acquire the knowledge and skills needed to promote sustainable development, including, among others, through education for

> sustainable development and sustainable lifestyles, human rights, gender equality, promotion of a culture of peace and non-violence, global citizenship, and appreciation of cultural diversity and of culture's contribution to sustainable development.
>
> (United Nations, 2015, p. 21)

Furthermore, according to the GAP, ESD:

- Allows every human being to acquire the knowledge, skills, attitudes, and values necessary to shape a sustainable future;
- Empowers learners to make informed decisions and responsible actions for environmental integrity, economic viability, and a just society, for present and future generations;
- Is a holistic and transformational education that addresses learning content and outcomes, pedagogy, and the learning environment;
- Achieves its purpose by transforming social institutions so they can respond creatively to global sustainability challenges.

(UNESCO Project Report, 2018, p. 6)

Acosta Castellanos, Queiruga-Dios, Hernández Encinas and Acosta (2020) suggest that ESD aims to develop a comprehensive transformation of educational systems to promote skills, knowledge and values and to equip people with the skills to make them able to co-create a sustainable future. However, Singer-Brodowski, von Seggern, Duveneck and Etzkorn (2020, p. 1) suggests that "the mainstreaming of ESD is becoming increasingly successful, questions of capturing and measuring the state of ESD within a certain region or sector are becoming more relevant". Manteaw (2012) alerts to the challenges concerning ESD implementation:

> The ESD discourse derives its appeal and popularity from the mother concept – sustainable development – a populous concept that imbues optimism and a sense of renewal; however, despite its growing popularity there still lacks clarity in the role of education in the quest for sustainable development, a situation which could well be traced to existing confusions surrounding the implication and application of sustainable development. These confusions have also made it difficult for most countries and their education systems to use the concept to develop a coherent philosophy to inform educational thinking and practice
>
> (Manteaw, 2012, p. 377)

Sustainability challenges in Southern Africa and elsewhere have a long history (Mandikonza & Lotz-Sisitka, 2016). Today, the earth systems and their functioning are being fundamentally altered by human activity, ushering in a completely new geological epoch, the Anthropocene (Steffen, Broadgate, Deutsch, Gaffney, & Ludwig, 2015). Asano (2011) suggests that ESD is a means for raising awareness and behaviour about the earth and is an important basis for lifelong learning and citizen participation, especially if done in early childhood. A comparative study of curricula in Sweden and Japan shows that ESD is present in the curricula for both countries (Fredriksson, Kusanagi, Gougoulakis, Matsuda, & Kitamura, 2020). However, it is implemented differently. Their research shows that in Sweden, ESD is mentioned as part of the principles that guide education, whilst in Japan, it is integrated into subject syllabi and emphasised. In another comparative study between England and Turkey, Bourn and Soysal (2021) found that ESD lacks prominence in teacher education courses offered in the two countries. Another study by Kalsoom, Qureshi and Khanam (2019), in Pakistan, found that teacher education documents did not incorporate the elements of sustainability education. As indicated to, the purpose of this chapter is to examine how ESD is reflected in science and technology teacher education policies and curricula in South Africa and Sweden. Additionally, in this chapter, we critically examine the nature of the science and technology teacher education programmes offered at different universities in South Africa and Sweden and their alignment to the policy and practice.

As previously noted, the UN declaration on SD forms the backdrop for this comparative study on science and technology teacher education policy and curricula between South Africa and Sweden. In Chapter 1, it was mentioned that this study is inspired by the collaborative research project involving the two countries. It provides an opportunity to examine how the countries responded to the UNs' call for ESD. Writing from Germany, Eilks (2015) states that educational domains and levels, including primary and secondary science education, have been working to contribute to education enabling younger generations to become responsible citizens and promote SD in our world. In South Africa, ESD is an educational approach that serves as a catalyst of social change and promotes sustainability (Teise & Le Roux, 2016). In Sweden, ESD is utilised as guide and approach to teaching. One example of this is the norm perspective approach to ESD wherein schools work towards building norm-supporting structures in the learning environment (Wickenberg, 2004).

## Methodology

As stated in the foregoing, in this chapter, we use document analysis to examine how the participating universities in South Africa and Sweden have approached ESD in their teacher education policy and linked science and technology-related school curricula. Akşan and Baki (2017) identify document analysis is an established research method used to make valid and reliable inferences from texts. Document analysis involves an analysis of content found in the documents to be analysed.

> "Content analysis has been defined as a systematic, replicable technique for compressing many words of text into fewer content categories based on explicit rules of coding. In its most simplistic form, it is a research tool used to determine the presence of certain words and concepts; a means of analysing texts" (O'Donoghue, Doody, & Cusack, 2011, p. 145).

Key concepts, terms and issues relating to ESD were used to develop an analytical tool to analyse the policy documents and university curricula for this study. The rationale for using this data collection strategy is that document analysis provides or supplements data, tracks change and development and suggests questions raised relating to a study (Bowen, 2009). The documents analysed included: the legislative teacher education policy, specific institutional teacher education curricula for different modules being taught at universities and some linked school science and technology curricula in South Africa and Sweden.

Integrating ESD in teacher education policy and curricula requires a change in mindset. However, change is not always easily embraced. As Læssøe, Schnack, Breiting and Rolls (2009) and McKeown (2002) suggest, lack of training and insufficient teacher knowledge are the major obstacles facing ESD. They also emphasised the need for effective teacher training. In view of this, this study had a specific focus on teacher education and training. Teacher education curricula from four different universities (two from each country) are analysed to identify the extent to which matters relating to SD and the Anthropocene are covered. One of the major aims of science and technology teacher education is to equip student teachers with the requisite knowledge and skills to teach content in the school curricula documents. With this in mind, our analysis of the teacher education policy documents and university and school-linked science and technology curricula was guided by the following questions:

- How do curriculum documents in South Africa and Sweden address issues of sustainability?
- In what ways do legislative and policy guidelines promote the teaching of science and technology for sustainability?
- How do these countries compare in addressing these issues in the Anthropocene?

The analysis of the teacher education policy documents, university curricula and school science and technology linked school curricula included attending to the following aspects:

- ESD target for each country;
- Philosophical and theoretically underpinning of the policies and curricula;
- Vision, programme design, curriculum and pedagogy, opportunities for practice
- Approaches for science and technology teaching; and
- Links between teacher education and practice.

The document analysis involves content analysis of the policies and linked science and technology curricula. A similar study by Kalsoom, Qureshi and Khanam (2019) used content analysis of teacher documents as a method of data collection in a study to examine the status of SD in teacher education in Pakistan. Their content analysis focused on ESD terms, content, processes and learning. In this study, an analytical tool, with agreed questions, was developed and content and construct validated. Ten different subject experts at the four participating universities (n = 10) were sent the analytical tool and conducted the analysis of the policy documents and university curricula. Their analysis of the policy documents and university curricula was the primary data used in the chapter. The tool drew from comparative studies frameworks such as the Conceptual Framework developed by Klette and Hammerness (2016) and the *Unit Based Sustainability Assessment Tool* by Togo and Lotz-Sisitka (2009). In developing the tool, cognisance was taken of the limitations of international comparative research in teacher education, identified by Afdal (2019). The tool takes into consideration whether creativity, innovation and "out-of-the-box" thinking are integral components of policy and curricula. Seatter and Ceulemans (2017) emphasise the need for today's students to be able to recognise the unsustainability of contemporary problem solving, cease searching for the "one right answer", and think instead in terms of good ideas and best solutions. The tool, therefore, goes beyond the conventional ticking sheeting and allows a detailed comparison of teacher education systems between the two countries.

Jenset, Klette and Hammerness (2018) have called for teacher preparation to be more deeply rooted in practice. Hence, our examination of the teacher education curricula, in the two countries, looked at the opportunities offered to student teachers to enact practice and content taught in science and technology modules. First, we analysed the teacher education policy documents for both countries. Thereafter, four modules or courses taught in science and technology teacher training programmes at the different universities were examined to identify aspects of sustainability. The last set of documents examined were school curricula in science and technology that student teachers must teach. For consistency, common subjects such as biology, which is taught in schools in different countries, were compared. This enabled us to determine the approach to ESD adopted by the different countries and decide whether it is a transformative or transmissive approach as suggested by Bourn and Soysal (2021). In this study, we examine whether ESD is treated as content to be taught (a body of knowledge) or is used as a mechanism to transform society. The analysis of university and linked school curricula also examines how the use of interdisciplinary teaching and multiple pedagogies in the different curricula unfold as the key characteristics of ESD.

## Results and discussion

### *Teacher education policy and linked curricula in South Africa and Sweden*

In presenting the results of the content analysis of the policies and curricula for the two countries, a narrative approach is adopted. In qualitative research, narrative analysis or inquiry focuses on the stories of people, their written or spoken words and experiences (Keats, 2009). However, in this chapter, the narrative analysis focuses on the written texts in policy documents and university curricula. Educational policies and university curricula are unpacked to determine how they address real-world problems such as sustainability. The content analysis of teacher education policy documents in both countries, modules and courses taught at universities and school science and technology curricula are examined to determine how both countries approach ESD. The narrative approach enables a deeper understanding of how ESD is adopted in both countries through the documents analysed.

### *Linked school curricula*

Both South Africa and Sweden have nine years of compulsory schooling. The current Swedish compulsory schooling curriculum is based on the 2010 Swedish Education Act (Skollagen) and focuses on pupils acquiring and developing knowledge and values. The policy document for compulsory schooling was revised in 2018. The advent of democracy in South Africa in 1994 ushered in many different curricula reforms such as Curriculum 2005 in 1997, National Curriculum Statement in 2002, Revised National Curriculum Statement in 2008 and Curriculum Assessment and Policy Statement (CAPS) in 2011 (Department of Basic Education, 2011). The CAPS is aligned to the National Education Policy Act, which was legislated in 1996. In both countries, compulsory schooling ranges from Grade R or preschool (six or seven years) to Grade 9 (15 or 16 years). The linked curricula examined in this chapter were teacher education curricula at universities in South Africa and Sweden and curricula taught in schools. Although aspects of sustainability are included in a wide range of curricula, the linked universities and schools' curricula examined in this chapter were science and technology-related. Table 2.1 shows some of the key principles and aims of the compulsory schooling curriculum in both countries.

In South Africa, the curriculum is prescribed and all schools have to follow it as set out in the CAPS for the different subjects. In contrast, in Sweden the core content (syllabus) is identified; however, schools are allowed the freedom to pursue the core content in their own creative way. The general sections of the Swedish compulsory school curriculum address sustainability (in its broadest sense) in many ways. Sustainability is addressed in three sections of the Swedish compulsory curriculum, section 1 (fundamental values and tasks of the school), section 2 (overall goals and guidelines) and section 5 (syllabuses). In South Africa, issues of sustainability are implicit in the curriculum. The aims of upper secondary schooling in Sweden differ from compulsory schooling, whereas in South Africa the aims of the curriculum do not change across the levels, from compulsory schooling to upper secondary schooling (or Further Education and Training, FET, in South Africa). From the principles and aims, it is clear that both countries value human rights, the focus of development of learners and promotion of 21st skills such as critical thinking, creativity, collaboration and communication. In addition, in both countries, there is respect for the environment. Hence, ESD is encouraged.

The introductory section of the Swedish curriculum for the compulsory school, preschool class and school-age educare focuses on several aspects of

*Table 2.1* Key Principles (Values) and Aims (Goals) in Preschool to Grade 9 Schooling in South Africa and Sweden Curricula

| | *South Africa* | *Sweden* |
|---|---|---|
| **Principles/ values** | • Social transformation<br>• Active and critical learning<br>• High knowledge and high skills:<br>• Progression<br>• Human rights, inclusivity, environmental and social justice<br>• Valuing indigenous knowledge systems<br>• Credibility, quality and efficiency | • Development and learning of all pupils<br>• Lifelong desire to learn<br>• Respect for human rights and the fundamental democratic values<br>• Respect for the intrinsic value of each person and the environment |
| **Aims/goals** | • Identify and solve problems and make decisions using critical and creative thinking<br>• Work effectively as individuals and with others as members of a team<br>• Organise and manage themselves and their activities responsibly and effectively<br>• Collect, analyse, organise and critically evaluate information<br>• Communicate effectively using visual, symbolic and/or language skills in various modes<br>• Use science and technology effectively and critically showing responsibility towards the environment and the health of others<br>• Demonstrate an understanding of the world as a set of related systems by recognising that problem-solving contexts do not exist in isolation | • Determine and express ethical standpoints based on knowledge of human rights and basic democratic values, as well as personal experiences<br>• Respect the intrinsic value of other people<br>• Reject the subjection of people to discrimination, oppression and victimisation and become involved in helping other people<br>• Empathise with and understand the situation other people are in and also develop the will to act with their best interests at heart<br>• Show respect and care for both the immediate environment, as well as the environment from a broader perspective |

Source: Adapted from Schooling Policy Documents in South Africa and Sweden.

sustainability by identifying values and skills needed for persons to act sustainably. The Swedish curriculum encourages a holistic perspective of sustainability (Swedish National Agency of Education, 2018). There is a *historical perspective*, which makes it possible to connect to aspects of the Anthropocene as it aims to create an understanding of "now" through "then" to make choices for "what is to come". There is an *environmental perspective*, relating to both local and global responsibilities, an *international perspective*, which relates to solidarity between people, and finally, an *ethical perspective*, which is to lead pupils being able to act responsibly in various situations. It is the teachers' responsibility to create opportunities for students to make meaning of "larger domains of knowledge". The Swedish compulsory curriculum caters to Biology, Physics, Chemistry and Technology, whereas Natural Sciences or Natural Sciences and

Technology are taught in South Africa. It is noteworthy that issues of sustainability are addressed specifically in the different subjects for both countries. However, some of the content in the Swedish curriculum is explicitly stated. For example, in Biology, some subcategories for pupils to study are Nature and Society, Body and Health, Biology and World Views or in Chemistry, Chemistry in Nature, Chemistry in Everyday Life and Society, Chemistry and World Views. However, in South Africa, the Natural Science knowledge strands (themes) are Life and Living, Matter and Materials, Energy and Change and Planet Earth and Beyond. The teachers in Sweden have the flexibility to develop the content for their subcategories. This latitude is not present in South Africa. Both South Africa and Sweden state the content to be taught in their policy documents.

The specific aims for all subjects taught in the compulsory schooling phase are found in the policy documents. In South Africa, the subject-specific aims are aligned to the general aims in the policy document, CAPS. The general aims are derived from the constitution of the country and serve to indicate the kind of citizen required in the country. In Sweden, subject aims and knowledge requirements in the syllabuses explain more on how students are to meet new knowledge and in what contexts this knowledge is to be developed. The subject description and core content are overarching and focus on "what to do". The Swedish curricula do not specify how teaching is to be carried out, but certain words (verbs) indicate activities that are more likely to provide opportunities to succeed, and these are echoed in the knowledge requirements.

### *Upper secondary or FET curricula*

Both South Africa and Sweden have three years of upper secondary schooling from Grades 10 to 12. In South Africa, upper secondary schooling is referred to as the FET phase. Learners who complete upper secondary schooling can register at a Higher Education Institution (HEI). Table 2.2 examines some of the key principles and aims of the upper schooling curriculum in both countries.

For post-compulsory schooling, with respect to their education, learners in both counties are free to pursue a vocational or academic route. Upper secondary schooling in both countries is discipline-specific. Subjects that pupils can pursue in both countries are Biology (Life Sciences), Chemistry and Physics (or Physical Sciences in the case of South Africa). The aims listed in Table 2.2 show that issues related to sustainability are embedded in the curricula for both countries. However, the countries have a different approach towards achieving their aims. In South Africa, all subjects target the same general aims as listed in Table 2.2. Individual subjects have specific aims that enable the general aims to be achieved. Life Science, for example, has three specific aims which teachers use to develop their learning objectives for the lessons they teach. For example, specific aim 3 (which is derived from the general aims above), in Life Sciences, highlights the need for learners to understand the applications of science in everyday life. In Sweden, all effort is directed towards achieving the aims outlined in Table 2.2. The Swedish curriculum for the upper secondary school (Swedish

*Table 2.2* Upper Secondary or FET Curricula

| | *South Africa* | *Sweden* |
|---|---|---|
| **Principles/ values** | • Social transformation<br>• Active and critical learning<br>• High knowledge and high skills<br>• Progression<br>• Human rights, inclusivity, environmental and social justice<br>• Valuing indigenous knowledge systems<br>• Credibility, quality and efficiency | • Development and learning of all pupils<br>• Lifelong desire to learn<br>• Respect for human rights and the fundamental democratic values<br>• Respect for the intrinsic value of each person and the environment |
| **Aims/goals** | • Identify and solve problems and make decisions using critical and creative thinking<br>• Work effectively as individuals and with others as members of a team<br>• Organise and manage themselves and their activities responsibly and effectively<br>• Collect, analyse, organise and critically evaluate information<br>• Communicate effectively using visual, symbolic and/or language skills in various modes<br>• Use science and technology effectively and critically showing responsibility towards the environment and the health of others<br>• Demonstrate an understanding of the world as a set of related systems by recognising that problem-solving contexts do not exist in isolation | • Acquire good knowledge in the courses they have studied<br>• Apply this knowledge for further studies and in societal life, working life and everyday life |

Source: Adapted from policy documents, CAPS Life Science and Sweden Curriculum for the Upper Secondary School.

National Agency for Education, 2018) highlights the importance of acquiring good knowledge and using the knowledge in societal and everyday life.

Table 2.3 uses a common subject in both countries, Life Sciences in South Africa and Biology in Sweden, to show how matters of sustainability are taught in upper secondary schooling. The Swedish upper secondary curricula identify an environmental perspective as a task of the school:

> *Environmental perspectives* in education should provide students with insights so that they can not only contribute to preventing harmful environmental effects but also develop a personal approach to overarching, global environmental issues. Education should illuminate how the functions of society and our ways of living and working can best be adapted to create sustainable development…
>
> (Swedish National Agency for Education, 2013, p. 6)

*Table 2.3* Applying Knowledge Upper Secondary or FET

| *South Africa* | *Sweden* |
|---|---|
| The following information is drawn from the Life Science CAPS document. By studying and learning about Life Sciences, learners will develop:<br><br>• Their knowledge of key biological concepts, processes, systems and theories<br>• An ability to critically evaluate and debate scientific issues and processes<br>• Greater awareness of how biotechnology and knowledge of Life Sciences have benefited humankind<br>• An understanding of how humans have impacted negatively on the environment and organisms living in it<br>• A deep appreciation of the unique diversity of past and present biomes in Southern Africa and the importance of conservation<br>• An awareness of what it means to be a responsible citizen in terms of the environment and lifestyle choices that they make<br>• An awareness of South African scientists' contributions<br>• Scientific skills and ways of thinking scientifically that enable them to see the flaws in pseudo-science in popular media<br>• A level of academic and scientific literacy that enables them to read, talk about and write and think about biological processes, concepts and investigations | This information is drawn from the Swedish upper secondary curriculum. Students:<br><br>• Should acquire good knowledge in the courses they have studied and can use this knowledge for further studies and in societal life, working life and everyday life<br>• Can use their knowledge as a tool to:<br>✓ Formulate, analyse, test assumptions and solve problems<br>✓ Reflect over their experiences and their individual ways of learning<br>✓ Critically examine and assess statements and relationships<br>✓ Solve practical problems and tasks |

Source: Adapted from Policy Documents, CAPS Life Science and Sweden Curriculum for the Upper Secondary School.

In South Africa, sustainability, as a topic, is not taught as a standalone subject. However, issues of sustainability are weaved into different aspects within the broad topic of Environmental Studies. It is clear from the above policy extracts that the curriculum aims to equip learners with skills ranging from cognitive, reading and writing (language), scientific thinking, hypothesising, critical thinking (science process skills), application to everyday life skills and practical skills (laboratory and field). The skills are vital in enabling learners to tackle sustainability issues. However, there are some important skills that are neglected, which include exercising creativity to solve problems and the use of technology in science.

An analysis of the policy document for Life Sciences in South Africa shows that the Environmental pillar of the ESD framework is evident in Grades 10–12 in the upper secondary (FET phase) even though it is left up to the teacher to plan and teach their lessons integrating ESD. Generally, the UN SDG goals are not explicitly stated but are implied in some topics in Life Sciences under the

*Table 2.4* Examples of Grade 11 SDG-Related Topics in the South African CAPS Document

| *Grade 11 caps topic* | *Its aligned sustainable development goal* |
|---|---|
| Roles of microorganisms in maintaining balance in the environment and web of life | Climate change: Take urgent action to combat climate change and its impacts |
| Symbiotic relationships: nitrogen-fixing bacteria in plants | Climate change: Take urgent action to combat climate change and its impact |
| Bryophytes (they thrive in moist environments) | Life on land: protect, restore and promote sustainable ecosystems |
| Loss of biodiversity | Life below water: conserve and sustainably use the oceans, seas and marine resources for development<br>Protect, restore and promote sustainable use of terrestrial ecosystems, sustainably manage forests, combat desertification and halt and reverse land degradation and halt biodiversity loss |
| The relationship between food intake, energy, growth and health requirements. A balanced diet and changing requirements with age, gender and activity levels | Good health and well-being: ensure healthy lives and promote well-being for all at all ages |

ecological dimension of the Life Sciences curriculum. As an example, sustainable development is highlighted when studying ecosystems. Table 2.4 shows how Life Sciences topics taught in South Africa link to the SDGs and ESD.

From Tables 2.3 and 2.4, it is clear that geological timescales form part of the curriculum and along with this, pupils are taught about the rapid loss of biodiversity due to anthropogenic activity – amongst other impacts covered in other topics. Hence, ESD is a crucial component of the FET or upper secondary school curriculum. The next section looks at the training of teachers to teach ESD.

### *Teacher education policy*

Higher education institutions (HEIs) play a critical role in transforming the mindsets of student teachers (Holm & Martinsen, 2015). Similarly, Lindström and Beach (2015) note:

> A country's teacher education is shaped by policies and procedures that are intended to enable access to the knowledge and skills that policy designers feel are appropriate for teachers to perform their school and classroom duties in the interests of pupils and the wider community.
>
> (Lindström & Beach, 2015. p. 248)

The nature of teacher education shapes the career of a student teacher. In Sweden, professional certification is required before a person can obtain a permanent post as a teacher. Whilst in South Africa, no certification is needed to assume a permanent post as a teacher. Any person who completes a teacher education qualification is entitled to teach. However, teachers are expected to register as members of a teacher professional organisation. In this section, we examine some of the basic competencies of a beginner teacher in South Africa and competencies of student teachers in Sweden. This provides insight into the outcomes, skills and values expected for student teachers. Teacher preparations programmes are constantly revised to accommodate societal changes. Teacher education policies in both South Africa and Sweden have been revised. In Sweden major revisions to the teacher, education policy took place due to political reforms and societal challenges. In Sweden, comprehensive reforms, resulting in a single integrated system in higher education, occurred in 1977 whilst a new act (Higher Education Act or Högskolelag) and ordinance (Higher Education Ordinance) was implemented in 1993. In 2007, the legislative structure for higher education was aligned to the Bologna process. Similar changes occurred in South Africa. In 1994, the newly elected democratic government in South Africa identified a policy framework to transform teacher education ensuring equity, redress and integration of all teaching qualifications. The first major revision took place in 2000, which focused on the Role of the Educator and was part of the Norms and Standards for Educators (DoE (Department of Education), 2000). The next major revision in 2011 focused on teacher identity by shifting from the Role of the Educator to the Basic Competencies of a Beginner Teacher. Minor revisions were made to this policy in 2015 (DHET, 2015). Table 2.5 shows the competencies needed in preparation for a teaching career.

Table 2.5 does not specifically mention anything about ESD. However, the South African legislative framework, Minimum Requirements for Teacher Qualification (MRTEQ), highlights the need for five kinds of learning, namely, disciplinary, pedagogical, practical, fundamental and situational. The three pillars for teacher training in Sweden focus on competencies in content, methodology and leadership. There is some synergy in the teacher education programmes in the two countries. Both countries stress the importance of content and pedagogical knowledge. In South Africa, there are fundamental skills such as literacy, numeracy and Information Technology (IT) that students must have before qualifying as a teacher. These skills are necessary due to the digital divide and multiple languages spoken in the country. In Sweden, leadership is valued and considered an essential competency for a teacher. In South Africa, this is not a competency but a role of a teacher. This role is merged with the roles of an administrator and a manager.

In Sweden, teacher education programmes focus on the teaching profession and the knowledge and skills future teachers need. The policy documents highlight critical perspectives on research, problem-solution, subject knowledge and methodology in the daily school practice. The policy documents in both countries do not focus on ESD or SD. There is no guidance on the relation of knowledge and sustainability. Knowledge is described as a means to prepare students for their future profession but mainly with a focus on subject knowledge needed and knowledge on generic teacher's needs, such as knowledge about how to evaluate and assess students' learning and skills.

*Table 2.5* Competencies of Student Teachers in HEIs

| *South Africa* | *Sweden* |
|---|---|
| The following are the minimum set of competencies required of newly qualified teachers:<br>• Newly qualified teachers must have sound subject knowledge<br>• Newly qualified teachers must know how to teach their subject(s) and how to select, determine the sequence and pace of content in accordance with both subject and learner needs<br>• Newly qualified teachers must know who their learners are and how they learn; they must understand their individual needs and tailor their teaching accordingly<br>• Newly qualified teachers must know how to communicate effectively in general, as well as in relation to their subject(s), to mediate learning<br>• Newly qualified teachers must have highly developed literacy, numeracy and Information Technology (IT) skills<br>• Newly qualified teachers must be knowledgeable about the school curriculum and be able to unpack its specialised content, as well as being able to use available resources appropriately, to plan and design suitable learning programmes<br>• Newly qualified teachers must understand diversity in the South African context to teach in a manner that includes all learners. They must also be able to identify learning or social problems and work in partnership with professional service providers to address these<br>• Newly qualified teachers must be able to manage classrooms effectively across diverse contexts to ensure a conducive learning environment<br>• Newly qualified teachers must be able to assess learners in reliable and varied ways, as well as being able to use the results of assessment to improve teaching and learning<br>• Newly qualified teachers must have a positive work ethic, display appropriate values and conduct themselves in a manner that befits, enhances and develops the teaching profession<br>• Newly qualified teachers must be able to reflect critically on their practice, in theoretically informed ways and in conjunction with their professional community of colleagues to constantly improve and adapt to evolving circumstances | Students within first and second cycle studies should develop:<br>• The ability to make independent and critical assessments<br>• The ability to identify, formulate and solve problems<br>• A preparedness to deal with changes in working life<br>In addition to acquiring knowledge and skills in the field covered by the course, students should develop the ability to:<br>• Seek and evaluate knowledge at a scientific level<br>• Follow developments in knowledge<br>• Exchange knowledge with people without special expertise in the field<br>Teacher programmes focus on three competencies:<br>• Content competence<br>• Methodological competence<br>• Leadership competence. |

Source: Adapted from the South African Minimum Requirements for Teacher Qualification Act and Swedish Higher Education Act.

As mentioned, the South Africa disciplinary learning (subject matter or content knowledge) is guided by the FET CAPS document considering that these student teachers are trained to teach compulsory and upper secondary (FET, Grades 10–12). The philosophical framework underpinning the MRTEQ is teachers are agents of change. Teachers are critical and transformational in everything they do. Active participatory and reflective learning shapes teacher education policy documents in both countries. Both policy documents aim to equip student teachers with the professional skills required to teach in the 21st century, respond to social inequalities and diversity. In the next section, we examine how the policy is implemented and whether aspects of ESD are addressed in modules offering at universities.

### *Module offering at universities*

For both South Africa and Sweden, module offerings at the different universities vary. However, all universities have a core (mandatory course), fundamental and elective (subject-specific courses) modules that student teachers need to take in their teacher preparation course. At both universities in Sweden, their teacher education programmes offer modules in Technology, Chemistry, Physics and Biology. Throughout the courses, issues of gender, subject-specific teaching and SD are treated. Some module outcomes focus on SD. Topics such as sustainable energy systems and ecosystems are taught in different modules. The exit level outcomes for teacher education programmes in Sweden require student teachers to be able to able to analyse, evaluate and provide a sound argument on matters relating to ESD. Modules taught at primary, upper primary and secondary teacher education programmes are linked to compulsory school syllabuses. The biology modules offered at the Swedish universities have an ESD component with ecology and ESD in the research field as subcomponents. However, most modules offered in Swedish universities do not offer ESD explicitly.

An identical scenario exists in South African Universities. Subject experts in the universities point out that sustainability is only taught as a topic within modules and not as a stand-alone module. Furthermore, if there are topics relating to sustainability in a specific module, research is encouraged. SD is treated as both a global and societal/contextual problem aligned to *situational* learning in MRTEQ. Some Natural Science teacher education modules teach about the Anthropocene as a component of the geological timescale. There are also many aspects in Natural Science modules that highlight the influence of human activity on the environment – ranging from abiotic factors to multiple levels of biological organisation. Student teachers are encouraged to take part in issues related to sustainability where they conduct sustainability projects. Sample projects conducted include the design of some clothes using only biodegradable and recyclable materials and with messages creating awareness about environmental issues. In the next section, we examine how both countries address issues in the Anthropocene.

### *Issues of the Anthropocene in school curricula policy documents*

In Sweden, the word Anthropocene is not present in any of the analysed curricula documents. However, there is a presence of the human aspect running as a common thread throughout the compulsory curriculum. The historical perspective that is repeated in the syllabuses enables pupils "to develop an understanding of the present, and a preparedness for the future, and develop their ability to think in dynamic terms" (Swedish National Agency for Education, 2018, p. 8). The Anthropocene is implied in this perspective. There is a similar trend in subjects like Biology, Chemistry and Physics. In Biology, for example, the historical aspect states:

> Teaching should create the conditions for pupils to be able to differentiate between scientific and other ways of depicting the world. Through teaching, pupils should get an insight into the world view of science with the theory of evolution as a foundation, and also get perspectives on how this has developed and what cultural impact it has had.
>
> (Swedish National Agency for Education, 2018, p. 166)

This implies that the Anthropocene is taught indirectly. Subjects such as Physics, Chemistry and Technology have similar statements, this means the Anthropocene is not taught in a deliberate way.

Analysis of the curricula policy documents in South Africa shows a similar trend to that of Sweden. The Anthropocene is not dealt with explicitly but is discussed in topics such as global warming and human impact on the environment in Life Sciences and Natural Sciences.

### *Anthropocene in teacher education policy documents*

In the South African and Swedish teacher education policy documents, there is no mention of the word Anthropocene either. Policy documents do not dictate the content to be taught at universities but identify the framework for teacher education by specifying the aims and principles. The Anthropocene is taught in some universities' modules where the course outline shows topics such as human's effect on nature and technological advancements.

## *Highlights from the comparisons between South Africa and Sweden*

The highlights are aligned with the research questions identified earlier.

Research Question 1 – *How do curriculum documents in South Africa and South Africa address issues of sustainability?*

- Issues of sustainability or ESD are not explicitly mentioned in the policy documents in both countries.

- Upper secondary schooling in both countries is discipline-specific. Issues of sustainability are addressed within specific topics in the different school subjects for both countries.
- The content to be taught in the South African curriculum is prescribed nationally whilst the schools dictate what is to be taught in the Swedish curriculum. Hence, teachers in Sweden have greater flexibility to teach matters related to sustainability.
- Both countries target 21st century skills for learners. This implies ethical citizens who respect the environment.
- In Sweden, the aims in upper secondary schooling differ from compulsory schooling whilst the aims do not change across the levels for schooling in South Africa. However, in both countries, curricula aim to highlight the importance of respect for the environment.
- The principles and aims in the South African curriculum are driven by the Constitution of the country, its historical past and the citizen it envisages. It envisages an enlightened citizen who supports sustainability.

Research Question 2 – *In what ways do legislative and policy guidelines promote the teaching of science and technology for sustainability?*

- Both countries offer content linked to ESD which is taught as topics within discipline-specific modules and not as stand-alone or integrated modules.
- The teacher education policy documents for both countries identify the teacher as a change agent and aims to equip student teachers with the professional skills required to teach in the 21st century. However, no unique pedagogical approach to ESD is championed to equip student teachers to be transformative.
- Leadership is a competency identified in the Swedish teacher education policy whilst in the South African teacher education policy, it is considered as one of the seven roles teachers must fulfil in the execution of their duties.

Research Question 3 – *How do these countries compare in addressing issues in the Anthropocene?*

- The term Anthropocene is not mentioned in school or university curricula in both countries.
- All policy documents in both countries ignore the term Anthropocene.
- There is a lack of integration and evidence of the manifestation of the Anthropocene in curricula materials.
- Subjects and modules taught in schools and universities teach topics where the Anthropocene is implied.

## Conclusion

The purpose of this chapter was to look at how ESD has been represented in teacher education policy documents and science and technology curricula documents in South Africa and Sweden. The comparison of the curricula and teacher education programmes in both countries shows that pupils and student teachers are provided with very little information on ESD. However, ESD is implemented in compulsory and upper secondary schooling as topics in certain subjects. For example, ecological issues form an integral part of both Biology (Life Sciences) teacher education and schooling subjects in both countries. Other topics emphasised are ecosystems, biodiversity, life cycles of plants and animals and life-supporting processes such as photosynthesis and respiration. ESD is not explicitly taught in these topics. ESD is treated as content to be taught separately in certain subjects like sciences and technology. In Sweden, ESD is mentioned as part of the general principles that guide education but in South Africa, it is implied as a responsibility. Both countries highlight the importance of respecting the environment in the education policy documents as their initiative towards ESD. However, in both countries, a holistic, integrated view of ESD is not seen in the policy documents. Both countries adopt a transmissive approach to ESD by focusing predominantly on environmental issues, rather than a transformative approach. Furthermore, neither country adopts an interdisciplinary or transdisciplinary approach to ESD. Both countries need to rethink their approaches to ESD so that it embraces a holistic view of the Anthropocene that encompasses social, environmental and economic challenges. A reform of current curricula is recommended.

The analysis of teacher training programmes in both countries shows a policy and practice dissonance and more needs to be done concerning ESD. Documents from teacher education, in both countries, do not provide evidence on what skills and competences student teachers ought to be equipped with to teach ESD. The articulation of what comprise these skills and competencies is yet to be formulated. This provides an opportunity for further research. Furthermore, the analysis reveals a lack of clarity on what are the specific knowledge areas, concepts or models, required to develop a holistic or integrated approach to sustainability in the education of student teachers. There is a need for supplementary perspectives, models and values that student teachers need to develop during their teacher education programmes so they acquire the necessary competencies and skills to be able to teach sustainability. For example, if interdisciplinarity is articulated, then this must be an explicit part of a teacher education programme.

The current status quo shows that matters relating to ESD are in the hands of teachers who need to plan the integration of ESD in their teaching. Teachers have to shoulder the responsibility of achieving the aims of the curricula without proper education in ESD. An area of further research is an examination of how teacher education policies and modules offerings at universities need to change to promote a unique pedagogical approach to ESD that is transformative and equips student teachers to be agents of social change as envisaged in the policy documents. Teacher education needs to prioritise ESD in both initial teacher education and for continuing professional development, including in-service courses. Placing ESD at the core of teacher education programmes requires a rethink on matters relating to, the curriculum, sustainability and a response to anthropogenic challenges. There is a need to make ESD an explicit science and technology education curriculum goal. ESD should not only be planned for and taught but also assessed. A rethink from all stakeholders in education is necessary so that the 2030 Agenda for SD can be achieved.

## References


Acosta Castellanos, P. M., Queiruga-Dios, A., Hernández Encinas, A., & Acosta, L. C. (2020). Environmental education in environmental engineering: Analysis of the situation in Colombia and Latin America. *Sustainability*, *12*(18), 7239.

Afdal, H. W. (2019). The promises and limitations of international comparative research on teacher education. *European Journal of Teacher Education*, *42*(2), 258–275. doi: 10.1080/02619768.2019.1566316

Akşan, E., & Baki, A. (2017). Content analysis of curriculum-related studies in Turkey between 2000 and 2014. *Educational Sciences: Theory & Practice*, *17*(3), 877–904.

Asano, Y. (2011). The comparative study of education for sustainable development in early childhood in Sweden and Japan: Through "The environmental epistemological model of 5 aspects". *Problems of Education in the 21st Century*, *32*, 24–32.

Bourn, D., & Soysal, N. (2021). Transformative learning and pedagogical approaches in education for sustainable development: Are initial teacher education programmes in England and Turkey ready for creating agents of change for sustainability? *Sustainability*, *13*, 8973.

Bowen, G. A. (2009). Document analysis as a qualitative research method. *Qualitative Research Journal*, *9*(2), 27–40. doi: 10.3316/QRJ0902027

Department of Basic Education [DBE]. (2011). *Curriculum assessment and policy statement (CAPS) life sciences*. Pretoria: Government Printers.

Department of Higher Education and Training. (2015). National Qualifications Framework Act (67/2008): Revised policy on the minimum requirements for teacher education qualifications. *Government Gazette* 596 (38487). Pretoria: Government Printers.

DoE (Department of Education). (2000). National Education Policy Act, 1996: Norms and standards for educators. *Government Gazette* 415 (20844). Pretoria: Government Printers. Retrieved July 16, 2020, from https://www.gov.za/sites/default/files/20844.pdf

Eilks, I. (2015). Science education and education for sustainable development – Justifications, models, practices, and perspectives. *Eurasia Journal of Mathematics, Science & Technology Education*, *11*(1), 149–158.

Fredriksson, U., Kusanagi, K. N., Gougoulakis, P., Matsuda, Y., & Kitamura, Y. (2020). Comparative study of curriculums for education for sustainable development (ESD) in Sweden and Japan. *Sustainability*, 12, 1123; doi: 10.3390/su12031123

Holm, C., & Martinsen, A. (2015). Mapping the relationship between higher education and sustainable development. *Studia Paedagogica*, *20*(4), 71–84.

IPCC. (2021). *Climate Change 2021: The Physical Science Basis. Contribution of Working Group I to the Sixth Assessment Report of the Intergovernmental Panel on Climate Change*. V. Masson-Delmotte, P. Zhai, A. Pirani, S. L. Connors, C. Péan, S. Berger, N. Caud, Y. Chen, L. Goldfarb, M. I. Gomis, M. Huang, K. Leitzell, E. Lonnoy, J. B. R. Matthews, T. K. Maycock, T. Waterfield, O. Yelekçi, R. Yu, & B. Zhou (Eds.), Cambridge: Cambridge University Press.

Jenset, I. S., Klette, K., & Hammerness, K. (2018). Grounding teacher education in practice around the world: An examination of teacher education coursework in teacher education programs in Finland, Norway, and the United States. *Journal of Teacher Education*, *69*(2), 184–197.

Kalsoom, Q., Qureshi, N., & Khanam, A. (2019). Teacher education for sustainable development in Pakistan: Content analysis of teacher education curriculum and standards. *Journal of Research and Reflections in Education*, *13*(1), 15–34.

Keats, P. A. (2009). Multiple text analysis in narrative research: Visual, written, and spoken stories of experience. *Qualitative Research*, *9*(2), 181–195.

Keeble, B. R. (1988). The Brundtland report: "Our common future". *Medicine and War*, *4*(1), 17–25.

Klette, K., & Hammerness, K. (2016). Conceptual framework for analyzing qualities in teacher education: Looking at features of teacher education from an international perspective. *Acta Didactica Norge*, *10*(2), 26–52. doi: 10.5617/adno.2646

Læssøe, J., Schnack, K., Breiting, S., & Rolls, S. (2009). *Climate change and sustainable development: The response from education. The recommendations from the International Alliance of Leading Education Institutes.* Retrieved June 10, 2019, from Dpu.dk: http://www.dpu.dk/Everest/Publications/Om%20DPU/Institutter/Institut%20for%20didaktik/20091211090537/Current- Version/Report%20with%20recommendations.final.dec09.pdf

Lindström, M. N., & Beach, D. (2015). Changes in teacher education in Sweden in the neo-liberal education age: Toward an occupation in itself or a profession for itself? *Education Inquiry*, *6*(3), 27020. doi: 10.3402/edui.v6.27020

Mandikonza, C., & Lotz-Sisitka, H. (2016). Emergence of environment and sustainability education (ESE) in teacher education contexts in southern Africa: A common good concern. *Educational Research for Social Change*, *5*(1), 107–130.

Manteaw, O. O. (2012). Education for sustainable development in Africa: The search for pedagogical logic. *International Journal of Educational Development*, *32*(3), 376–383.

McKeown, R. (2002). *Education for Sustainable Development Toolkit.* Retrieved June 10, 2019, from Esdtoolkit.org: http://www,esdtoolkit.org/esd_toolkit_v2.pdf

O'Donoghue, G., Doody, C., & Cusack, T. (2011). Physical activity and exercise promotion and prescription in undergraduate physiotherapy education: Content analysis of Irish curricula. *Physiotherapy*, *97*(2), 145–153.

Sachs, J. D., Kroll, C., Lafortune, F., Fuller, G., & Woelm, F. (2021). *Sustainable development report 2021.* Cambridge: Cambridge University Press.

Seatter, C. S., & Ceulemans, K. (2017). Teaching sustainability in higher education: Pedagogical styles that make a difference. *Canadian Journal of Higher Education*, *47*(2), 47–70.

Singer-Brodowski, M., von Seggern, J., Duveneck, A., & Etzkorn, N. (2020). Moving (reflexively within) structures. The governance of education for sustainable development in Germany. *Sustainability*, *12*(7), 2778.

South Africa National Department of Education. (1997). *Curriculum 2005: Lifelong learning for the 21st century.* Pretoria: National Department of Education.

Steffen, W., Broadgate, W., Deutsch, L., Gaffney, O., & Ludwig, C. (2015). The trajectory of the Anthropocene: The great acceleration. *The Anthropocene Review*, *2*(1), 81–98.

Swedish National Agency for Education. (2013). Curriculum for the upper secondary school. Stockholm: Swedish National Agency for Education.

Swedish National Agency for Education. (2018). *Curriculum for the compulsory school, preschool class, and school-age educare: Revised 2018.* Stockholm: Swedish National Agency for Education.

Teise, K., & Le Roux, A. (2016). Education for sustainable development in South Africa: A model case scenario. *Africa Education Review*, *13*(3–4), 65–79. doi: 10.1080/18146627.2016.1224584

Togo, M., & Lotz-Sisitka, H. (2009). *Unit-based sustainability assessment tool. A resource book to complement the UNEP mainstreaming environment and sustainability in African universities partnership.* Howick: Share-Net.

UNESCO. (2009a). *Bonn declaration.* Retrieved from http://unesdoc.unesco.org/images/0018/001887/188799e.pdf

UNESCO. (2009b). *United Nations decade of education for sustainable development (DESD, 2005-2014), review of contexts and structures for education for sustainable development 2009.* Paris: UNESCO.

UNESCO. (2012). *Education for sustainable development, sourcebook.* Paris: UNESCO.

UNESCO. (2013). https://en.unesco.org/themes/education-sustainable-development

UNESCO Project Report. (2018). *Sustainability starts with teachers: Project report.*

United Nations. (2015). *Transforming our world: The 2030 Agenda for sustainable development.*

Weldemariam, K., Boyd, D., Hirst, N. et al. (2017). A critical analysis of concepts associated with sustainability in early childhood curriculum frameworks across five national contexts. *International Journal of Early Childhood, 49,* 333. doi: 10.1007/s13158-017-0202-8

Wickenberg, P. (2004). Norm supporting structures in environmental education and education for sustainable development. In P. Wickenberg et al. (Eds.), *Learning to change our world? Swedish research on education & sustainable development.* Lund, Sweden: Studentlitteratur.

# 3 (Re)thinking teacher education in the Anthropocene

## Perspectives from South Africa and Sweden

*Kassahun Weldemariam, Nina Christenson, Alexina Thorén Williams, Emmanuel Mushayikwa and Suresh Kamar Singh*

## Introduction

As humanity occupies the earth in the aftermath of the industrial revolution, it is widely agreed that we have entered a new geological era referred to as the Anthropocene. Scholars argue that education has a key role to play in responding to Anthropocene challenges. Yet, studies (e.g., Kopnina, 2014) argue that education practice can have a counterproductive and counterintuitive effect by unintentionally fostering an exclusively instrumental ideology of economic development, innovation as technical solution and growth eclipsing the bigger existential questions, such as asking questions about who we are and what we are doing to the planet. This includes the possibility of sustaining the environment in which living can take place and that growth can be about so much more than mere production. Overall, what forms of education and learning are sought in the Anthropocene remains obscured. What kind of education is responsive to the knowledge, values, ethics and responsibilities that the Anthropocene calls for? Such a question is, all too often, not even asked nor contemplated. It is an aporia at the heart of practice including the practice of teacher education (TE). Teacher education is a potentially well-situated programme to ask such questions as it opens up possibilities and pave the way to reach out to students in school settings.

In this chapter, we investigate four selected TE programmes in South Africa and Sweden against the Anthropocene predicament. Our aim is to compare and explore how these TE programmes are responding to the challenges posed by the Anthropocene. The comparative analysis is anchored around these four aspects: essence of the programme, views of knowledge, goals and learning outcomes and core competence emphasised. The chapter begins with an introduction to the essence of the Anthropocene and the role education plays in this new geological epoch. This is followed by a description of education in today's neoliberal economy and pertinent critique around it. Later, drawing on existing literature, the kind of knowledge, skills and values that may be sought in the

DOI: 10.4324/9781003190158-5

Anthropocene are identified and presented followed by a discussion of how TE responds to Anthropocenic challenges.

## Background

It is widely agreed that humanity is facing a range of sustainability issues such as: rising inequity, loss of democracy, runaway climate change and mass extinction, which are referred as characteristic features of the Anthropocene. The term Anthropocene highlights the human exceptionalism scenario wherein human consumptive and destructive activities have increasingly and widely altered the planet's ability to regulate and sustain itself (Crutzen & Stoermer, 2000). The Intergovernmental Panel on Climate Change (IPCC, 2021) has clearly indicated the precarious state of the planet and indicated that we are in a rapidly dwindling critical window to change: such that it is already an emergency, even a crisis and rapidly becoming now or never. There is an urgent imperative to alter our way of being in the world, to find an alternative way of knowing and a new ethical value: otherwise we, jeopardise our own existence as a species. Gibson, Rose and Fincher (2015) argue that to reverse or change the damaging human-centric behaviours, we first need to change our way of thinking and become aware of the scenario. From the vantage point of sustainability, ontological perspectives seeing the world as an object and us as an enquirer and knowledge-maker discrete and separated from that world are dysfunctional and potentially dangerous (Weldemariam, 2020).

Education plays an important and pivotal role in dealing with the challenges in our changing planet. The central role of education for sustainability has been emphasised since the introduction of Agenda 21 (Lafferty & Eckerberg, 2013). More recently, UNESCO's (2016) global education monitoring report has also emphasised the potential and critical role that education plays in connecting people and the planet. Likewise, there have been several global initiatives within education in order to address global environmental challenges. One significant initiative was the establishment of United Nations Decade of Education for Sustainable Development (2005–2014) aimed at integrating the principles, values and practices of sustainable development (SD) into all aspects of education and learning. However, education for sustainable development (ESD) has been criticised for its narrow perspective and human-centric characteristics which remain short of challenging anthropocentrism, thus making it difficult to solve the current environmental problem and reach sustainability (Weldemariam, 2020). Highlighting its limitations, Kopnina (2014) pointed out that "ESD masks its anthropocentric agenda and may in fact be counterproductive to the efficacy of environmental education in fostering a citizenry that is prepared to address the anthropogenic causes of environmental problems" (p. 73). Following the ESD decade, the global action programme, GAP, 2015–2019, has been introduced (United Nations Educational, Scientific and Cultural Organization (UNESCO), 2014).

Despite these initiatives, education has not succeeded in providing learners the required level of critical thinking to examine the structures and systems of beliefs that allow and promote unsustainable livings and thus concomitantly has continued to contribute to unsustainability (UNECE, 2012). Even higher education, a sector construed as having a key role in the transformative process towards sustainable societies and through their functions of education, research and outreach (Fadeeva & Mochizuki, 2010), is counterproductively contributing to the sustainability crisis (Tilbury, 2011). This scenario begs the question: is the type of education in today's higher education appropriate to enable students with the knowledge, skills and values needed to deal with the "wickedness" of Anthropocene challenges? Orr (1990) claims that more education of the same type will only make the problems worse: a type of education that "emphasizes theories, not values; abstraction rather than consciousness; neat answers instead of questions; and technical efficiency over conscience" (p. 238). Wiek, Withycombe and Redman (2011) has called for the need to pay attention to the role of education in enabling students with the abilities to be aware of their own (and dominant) assumptions and worldviews; to evaluate them from critical, systemic and futures' thinking and to shifting them. In parallel, Merino de Diego, and Díaz Carmona (2018) argue that "such type of work on the students' awareness and subjectivity is a necessary condition to debunk the myth of growth, as a necessary condition for living within planetary boundaries, and within the logics of social justice and human (and non-human) well-being" (p. 2).

## Critique around education in today's neoliberal economy

Scholars such as de Diego and Díaz Carmona (2018) argue that today's education is mainly serving the economy and is mainly aimed at making students employable and "uncritical consumers". They contend that "the currently dominant paradigm of education, which is transmissive and largely designed to serve the market needs, is at odds with the 'wickedness' problems of this new era of the Anthropocene built on acceleration and growth" (de Diego & Díaz Carmona, 2018, p. 3). Likewise, they highlight that "wicked" problems of the Anthropocene cannot be solved with the conventional linear and fragmented thinking, but rather by thinking through systemic relationships (de Diego & Díaz Carmona, 2018). Building on their argument, Assadourian (2017) points out that "Schooling today tends to ignore the massive changes that are looming and offers little in the form of preparation for slowing those changes or coping with them" (p. 6).

Owing to the aforementioned critiques, Patrick (2013) highlighted the need to resist neoliberal knowledge economy and instead suggested to consider emancipatory education that transforms the self-engendering human. He suggests that higher education institutions should strengthen practices that enable individuals to develop in ways that are consonant with "their sense of their own existence" (Patrick, 2013). Perhaps this sense of existence is "being a consumer" and is emanated from the idea of a society that is oriented around the notion to grow economically. In some consumer society, it is not uncommon to encounter scenarios wherein education, unintentionally, plays a maladapted role in society.

One instance of such a scenario is the commodification of knowledge opening up rooms for corporations to influence school curriculum, thus leading to the students to be indoctrinated into becoming uncritical consumers. Thus, there may be resistance against the sedimented hegemonic assumptions and hence the transformation needed and that this resistance goes beyond school and teachers. This may bring about a pertinent question: how can TE prepare future teachers for such societal and economic forces?

Another area of concern is the way that knowledge and coming to know is understood. Brennan (2017) indicates that today's education constitutes knowledge as abstraction, decontextualised from place, relationships and action, and perceived as past rather or prospective and future-invested. This entails that knowledge is viewed as mostly private and individually "owned" rather than understood as co-created in action. Additionally, there is a tendency to prescribe and proscribe for teachers what should be taught and how. School education today is almost closed ended, and the learning content is predetermined (Chawla & Cushing, 2007), thus engendering teacher hostility to critical approaches (Barrett & Sutter, 2006).

Others criticise today's education for its anthropocentric characteristic, and hence reinforcing a human nature binary where the human is seen as a subject outside of nature as a static object rather than a complex entanglement where human actions feed back into nature and change the whole (Malone, 2017; Taylor, 2013). This division between human and nature, where values concerning nature/the biosphere are treated as an add-on to human centric values is expressed in both the curriculum for compulsory school (Swedish National Agency for Education, 2019) and the TE study plans (Karlstad University, 2011; University of Gothenburg, 2020). These critiques around modern education beg pertinent questions of how today's schools are actually preparing students, as change agents – to face and deal with the social and ecological reality in the Anthropocene? Are teachers being prepared accordingly?

## Knowledge and values that might be sought in the Anthropocene

The Anthropocene calls for education to evolve towards a more emancipatory approach that enable students to become critically aware of how they perceive the world, with a view to fostering citizen engagement with socio-environmental issues (Jickling & Wals, 2008; Wals & Jickling, 2002). This entails that students are to be reflective of their own learning and their values, challenge hegemonic social beliefs and be able to navigate through terrains of complexity and uncertainty. Among other qualities, this requires moral skill, which is often left out in sustainability education which often addresses the development of normative competence (Wiek et al., 2011). In most cases, discussion around sustainability education tends to privilege information-processing and problem-solving competences at the expense of moral skills and self-discovery (Adomßent et al., 2014; Hesselbarth & Schaltegger, 2014; Rieckmann, 2012).

For instance, virtue education/character education focuses on the development of our moral character rather than the obedience of a prescriptive principle or rule, which helps to promote a broadly shared moral framework (Crossan, Mazutis, & Seijts, 2013). Focus on virtues enables learners to understand themselves better, to challenge their own assumptions and to acquire new perspectives. Furthermore, as virtues/strengths are moral dispositions that orient our judgement and behaviour towards the good, they may help to understand (and feel) sustainability and post-growth societies as an attitude or a way of life, rather than as a list of predefined practices to be performed.

While introducing the concept of earth education, Assadourian (2017) pointed out: "for humanity to get through the coming century, our schools must emphasize six new set of proficiencies which includes: earth-dependence, interdependence, creativity, deep learning, earth-centric leadership and life skills" (p. 8). He identified these principles as "a Common Core-equivalent that will enable us to survive life on a changing planet" (p. 8)" This further strengthens the need to rethink education for our challenges and responsibilities in the 21st century. Likewise, Sterling and Jickling (2017) highlighted that criticality and urgency are important concepts to embedded into education, which requires new frames of learning such as post-humanism, post-materialism, post-sustainability, etc.

Drawing on the aforementioned scholars, one could infer that the competencies required in the Anthropocene include: dealing with complexities, connections and relations, interdependence, ambiguity, anticipatory thinking, sensitivity and responsiveness, emotional intelligence, moral agency, empathy, sympathy, critical thinking and action competence. However, it is evident that these competencies are taught in today's TE? If yes, to what extent? If not, it is pertinent to ask how and where do teachers learn to do all the things they need to be able to do. Sund and Öhman (2014) assert that environmental and sustainability education (ESE) cannot be added to education but rather be seen as part of the pedagogies/didactics teachers enact. Furthermore, in addition to the three dimensions of sustainability (natural environment, society, economy), education in and for a sustainable world must include a political and moral dimension (Sund & Öhman, 2014). Thus, teachers need pedagogically/didactically relate to both the universal values of SD, global justice and the intrinsic value of nature, articulated in policy documents and the particular values linked to local communities, traditions and living conditions. Values that people live through in their everyday lives. However, navigating in the tension between these value dimensions requires the competencies highlighted above.

This further necessitates the need to examine the knowledge, values and skills that current TE strengthens and weakens. How responsive are our TE programmes to the Anthropocene in our institutions? It is worthwhile to explore priorities for pre-service teachers to learn in contemporary Anthropocene era. The knowledge we emphasise and appreciate in today's TE programme needs to be examined.

## Teacher education in the Anthropocene

Teacher education plays a key role for a sustainable world. Unlike other forms of education, TE is uniquely relevant as it addresses teachers embarking on to teach the future young generation who are the future change agents in school settings. Thus, it is central to explore how TE practice can be transformed so that it can respond to and address Anthropocene challenges.

In recent years, there has been discussion (Brennan, 2019) on the need to reorient education and TE towards SD. These discussions elucidate that TE is no longer about merely mastery of content knowledge and skill in a particular discipline that one is trained for. It rather necessitates the need for the development of key competences, values and skills needed to deal with global challenges of our time. Yet, it is not well documented how TE programmes in different nations have successfully transformed and integrated SD into the core of TE programmes.

While the integration of SD into TE is still under question, the contemporary Anthropocene era adds another layer of complexity to the problem. The notion of Anthropocene places the human in a new position (entangled with nature) and calls for a rethinking of knowledge, skills and values that are required to live in this new era. As indicated in the earlier sections, many scholars argue that the knowledge structure in modern TE is anthropocentric and hence built on a humans-nature binary. In doing so, the concept on Anthropocene urges us to question the implicit division between society and nature embedded within SD discourse and hence calls for rethinking its meaning (Johnson & Villumsen, 2018) and attributed ethical implications.

Hence, the Anthropocene poses challenges for education which require significant changes of practice in schools and universities, and thus to TE. Yet, the relationship between the Anthropocene and TE remains an underexplored topic. Drawing on empirical cases from four different TE programmes in South Africa and Sweden, this chapter addresses the following questions. How TE can respond/is responding to the challenges posed by the Anthropocene? What can TE contribute in terms of providing directions to rethink humans' relationship with the planet in today's Anthropocene era? What kind of (fresh) approach to TE is required in the Anthropocene/in 21st century?

## Theoretical framework

The notion of critical inquiry (Battersby & Bailin, 2011) has guided the comparative analysis among the four-teacher education programmes. This approach is chosen as it permits the opportunity to engage in relevant and context-related critical thinking. The authors followed the main stages of critical inquiry, which include: reading steering documents regulating the TE programmes, reconstructing the main arguments related to the topic and responding to the claims made (Boylan, 2009). According to Battersby and Bailin (2011), critical inquiry is "a dialectical process involving the comparative weighing of a variety of contending positions and arguments-while argumentation is seen as a way of arriving at reasoned judgments on complex issues" (p. 1.). Furthermore, Battersby and Bailin (2011) highlighted that there are six aspects of critical inquiry which include: the dialectical context and current debate around an issue; state of practice and beliefs surrounding an issue; political, historical and social contexts; knowledge of the relevant disciplinary context; sources of an argument, and awareness of one's own beliefs and biases. Adopting these procedures, the authors examined and compared the four-teacher education programmes in light of the Anthropocene.

## Methodology

Methodologically, we employed a comparative analysis of four-teacher education programmes based on collaborative inquiry. The investigation was undertaken by members of a research collaboration between South Africa and Sweden during the work with this book project. We anticipated that such an international collaboration would "demystify and elucidate the process of constructing knowledge" (Bray, Lee, Smith, & York's, 2000, p. 19) pertaining to TE in connection to contemporary Anthropocene era. This collaboration emerged in 2017 to discuss issues of TE within science and technology education in the two nations.

Members continued to explore and broaden this theme by collecting data within a table for each of the four comparative aspects. The tables and the ensuing bodies of text were scripted by each authors. This community of practice (Lave & Wenger, 1991) was further strengthened by face-to-face workshop meetings both in Gothenburg and Johannesburg in 2019. The primary description and investigation of each TE programme was conducted by each author from their home institution, drawing on local context knowledge and experience, thus enhancing the grounding of the investigation. In addition, the authors shared the "coding frame" (Silverman, 2011, p. 65) during the data collection and analytic processes through the completion of evolving tables with the commonly identified aspects. The analysis of each TE programme focuses on the following four aspects: description and discussion of the overall essence of each TE programme, how knowledge is viewed in each programme, each programme's goals/intended learning outcomes and emphasised core competence. Focusing on the aforementioned four coding themes identified, an inductive thematic analysis was employed during the research process (Guest, Namey, & Mitchell, 2012). As the process developed, the interplay between individual and group reflection was conducted through dialogues (Bray et al., 2000) which sometimes led to a rethinking of our own knowledge and understanding of our respective institution's TE programme.

While structuring our analytical framework and analysing the TE programmes, we employed the following procedure. First each author began by thoroughly scrutinising his or her institution's TE programme. This was followed by successive discussion among the authors team which resulted in the identification of the aforementioned four comparative aspects to focus and guide the comparative analysis. Once the aspects were identified, each author re-examined his or her respective TE programme in the light of what is required in the Anthropocene. Subsequently, the four aspects in the four teacher education programmes became the focus of analysis and discussed in light of the knowledge, skills and values that need to be taught in the Anthropocene. The overriding goals in the four programmes will serve as a point of departure for the analysis and discussion. For the purpose of meaningful exemplification and elucidation of each TE programme, excerpts are quoted in the comparative tables.

### *Description of the four teacher education programmes*

The following sections present brief description of the four teacher education programmes compared.

#### *Primary school TE programme at Gothenburg University*

The primary TE at Gothenburg University offers three programmes aimed at preparing student teachers for lower primary school (students age 6–9), upper primary school (students age 10–12) and primary school towards after-school centre/school age educare (students age 6–12). All three programmes are four-year-long but differ in terms of subject courses. For example, the programme for lower primary school prepares student teachers in the school subjects; Swedish, Mathematics, English, Natural Science, Technology and Social Science. In the upper primary programme, and in addition to reading the compulsory courses Swedish, English and Mathematics, the student teachers choose an enrollment that either gives them qualifications to teach Natural Science or Social Science subjects, or an enrollment that qualifies them to teach art, sports and health or music. In the programme directed towards working in the after school centre/school age educare, an additional subject is included besides core course in educational science (i.e., Sports and Health, Music, Home and Consumer Studies).

In the Education Plans for the TE programmes directed towards pre-school class and primary school grades 1–3, and towards primary school grades 4–6, it is stated that the "student teacher should be able to make decisions in the pedagogical work based on relevant scientific, social and ethical aspects, with particular consideration of human rights, and especially in the rights of the child according to the Convention on the Rights of the Child, and sustainable development" (p. 5). It is only in this formulation throughout the education plan that the concept of SD is explicitly addressed. On the other hand, several skills and abilities are highlighted which are, to a large extent, similar to the indicators of sustainable development goal 4, (SDG 4), Quality education for all (United Nations Educational, Scientific and Cultural Organisation (UNESCO), 2016). For example, to critically and independently practice, systematise and reflect on their own and others' experiences and relevant research results; to demonstrate in-depth ability to create the conditions for all pupils to learn and develop; to demonstrate the ability to communicate and anchor the school's core values, including human rights and fundamental democratic values; to demonstrate the ability to prevent and counteract discrimination and other offensive treatment of students; to show the ability to consider, communicate and anchor gender equality and equality perspective in pedagogical activities.

### *Primary school TE programme at the University of Limpopo*

The University of Limpopo (UL) offers three initial TE programmes. The first two programmes are four-year programmes and are phase related, namely, the Bachelor of Education for Foundation Phase teaching and Bachelor of Education for Senior and Further Education and Education Phase teaching. These programmes prepare teachers for grades 1–3 and for grades 7–12 in the schools. The other programme, a Post Graduate Certificate in Education, is a one-year programme and caters for students who already have an undergraduate degree and would like to attain a teaching qualification. All TE programmes align to the legislative policy for TE which is the Minimum Requirements for Teacher Qualification (MRTEQ). The policy documents do not dedicate the content that must be taught in the programmes but identifies the kinds of knowledge and competencies that pre-service teachers should develop.

### *Primary school TE programme at Karlstad University*

Reviewing the teacher programme at Karlstad University for primary education, grades 4–6, it is clear that SD perspective is relatively absent. In the study plan for TE for primary school (years 0–3 and 4–6), one of the learning outcome requirements is related to SD stating that students should be able to demonstrate the ability to make assessments of educational processes on the basis of relevant scientific, social and ethical aspects with special consideration of human rights, especially children's rights according to the Convention on the Rights of the Child, and SD.

Moreover, in a review of all the steering documents for all the courses included in the year 4–6 programme at Karlstad University, it is shown that it is only in semester 4 (out of 8) SD is included. During the fourth semester, the students can choose to take one out of four specialisations (30 credits); music, arts, social science oriented subjects or natural science oriented subjects. In music, SD can be found in one of the courses objectives as well as in one of the social science subjects – geography. In the course regarding the natural science subjects SD can be found within the objectives in biology.

Due to the aforementioned very limited focus on SD in TE at Karlstad University, the TE board recently decided on a new policy regarding these issues. According to this policy, all subjects that are included and compassing at least 30 credits should include at least one learning objective related to SD. However, due to the fact that the teacher students do not take any subject that exceeds the limit of 30 credits, the main responsibility for including these perspectives is given to the general education core courses which are courses common for all students of the programme regardless of their subject specialisation. The conclusion is that the focus on SD will increase but a concern remains about how the courses and subjects will have the opportunity to coordinate their work. In addition, no time schedule of the implementation process of the new policy has not yet been decided on.

### *Teacher education programme at the University of the Witwatersrand*

The primary school teacher preparation programme at WITS University follows the same guidelines as set for the Higher Education Institutions – teacher preparation programmes – i.e., four years, of which one year is teaching experience (teaching practice in schools), distributed within the four years (Department of Higher Education and Training, 2015). This means that during the period of their preparation, both primary and secondary teachers-in-education spend at least 6 weeks a year on teaching experience. Although the policy documents do not specifically refer to sustainability aspects, sustainability is built into the programme through its structure.

The WITS School of Education provides five initial teacher education (ITE) programmes. These are four-year Bachelor of Education studies in Foundation Phase, Senior Primary and Further Education and Education (FET) phase programmes and one-year Postgraduate Certificate in Education (PGCE) studies in Primary and FET studies. All these programmes however are embedded and integrated to ensure that all our graduates leave the institution with the same attributes which emphasise adaptability and sensitivity to the situational context in which the teachers find themselves.

Student teachers' reading for the four-year bachelor of education (BEd) programme has to cover the core competencies, as well as the subject matter specialisms. The breadth of the Initial Teacher Education (ITE) curriculum for both the BEd and the PGCE streams is aimed at producing graduates who are immersed in the teaching profession, graduates who are adaptive and versatile. The two bachelor of arts in teaching (BAT) programmes (full year courses) are meant to develop reflective practitioners and lifelong learners.

The WITS School of Education has positioned itself as a research-led and research-based TE institution. It is unique in that it is a self-contained institution and provides the disciplinary and pedagogical content without being dependent on the "home disciplines" (e.g., Psychology, Philosophy, Sociology, Math and Science among others). Though, not a Faculty, it has the capacity of one and can provide total immersion in educational professional learning. This deliberate organisation of the WITS School of Education enables sustainability by capacitating the school to manage change, as it is fine-tuned to the situational context of the teaching profession within the country.

Sustainability is also indirectly captured in the school's vision which emphasises graduate attributes of openness and readiness to adapt to change and to change with the context. The school's policy describes the professional identity of its graduates as:

a Reflective and enquiry driven informed practitioners
b Adaptive to the situational context of their school
c Possessing high levels of subject content knowledge
d Possessing high levels of pedagogical content knowledge, both in terms of content and context

It is arguable that these attributes of professional identity enshrine elements of SD in TE in as much as they emphasise the practice of education within a situational context. The emphasis on awareness of the South African context ensures that all our graduates are familiar with the South African historical past and the transformative imperatives facing the nation for its SD.

Kimathi and Rusznyak (2018) observe that the core competencies of BAT 1 and 2 emphasise that the student teachers should acquire intimate knowledge of South African history and transformation imperatives. These include human rights, social and economic SD, equality and the teacher's roles and responsibility in bringing about transformation in the country. The rest of the competencies in the core cluster are aimed at providing these student teachers with holistic skills necessary to bring about social transformation and SD in the South African context.

We will use Tables 3.1–3.4 to compare the four-teacher education programmes in light of the identified four aspects used for comparative purpose.

Three of the TE programmes have an explicit focus on content knowledge and the Swedish programmes emphasise pedagogical leadership. The TE programmes in South Africa mainly focus on situational contexts and experiences and the WIT programme also states that TE should encompass transformative practices. WIT and GU programmes stress importance of the TE to be research driven.

*Table 3.1* Essence of the Programme

| *Gothenburg University (GU)* | *University of Limpopo (LIM)* | *Karlstad University (KAU)* | *University of the Witwatersrand (WIT)* |
|---|---|---|---|
| Focus on specialisation in subject and subject didactic studies, knowledge as means to increase teaching quality<br>Research and experience-based education where pedagogical leadership, assessment, strengthened scientific knowledge and special pedagogical competence are important | Focus on specialised content knowledge, educational theory, teaching methodology, teaching skills and contextual and situated experiences | Focus on content competence, methodological competence and leadership competence | Focus on research-driven, research-led, transformative and situational context-embedded TE |

*Table 3.2* View of Knowledge

| *Gothenburg University (GU)* | *University of Limpopo (LIM)* | *Karlstad University (KAU)* | *University of the Witwatersrand (WIT)* |
|---|---|---|---|
| Knowledge as knowing about, literacy<br>Knowledge for fostering certain behaviour<br>Knowledge as a value approach<br>The ability to identify need for future knowledge<br>Develop knowledge and skills in pedagogical practices<br>Knowledge as a need, both learners (pupils) but also own knowledge for continuous development as teachers | PCK-focus:<br>Subject matter content knowledge<br>Pedagogical content knowledge<br>Curricular knowledge<br>Knowledge as vehicle for:<br>Rural regeneration, empowerment, development and caring<br>Lifelong learning, reflection, transformation and interdisciplinary approach<br>Knowledge is constantly evolving and a tool for empowerment, social justice and transformation | Knowledge as knowing about, literacy<br>Knowledge for fostering certain behaviour<br>Knowledge as a value approach<br>The ability to identify need for future knowledge<br>Develop knowledge and skills in pedagogical practices<br>Knowledge as a need, both leaners (pupils) but also own knowledge for continuous development as teachers | Focus on knowledge as a vehicle for social and economic transformation<br>PCK-focus:<br>Subject matter content knowledge<br>Pedagogical content knowledge<br>Curricular knowledge |

The two Swedish TE programmes' view on knowledge is identical due to the fact that these programmes are under the jurisdiction of the Swedish Ministry of Education and Research. Both TE in South Africa have a shared view on knowledge as a tool or vehicle for social (both), economic (WIT) justice. Hence, knowledge is seen as a tool for empowerment. Overall, subject content knowledge and knowledge of general and subject didactics are put forward in the South Africa TE and in Swedish TE pedagogical content knowledge (PCK) (Shulman, 1986) framework is highlighted. This mirrors the different tradition of using didactics or PCK as frameworks for teacher competences but here PCK and didactics can probably be seen as comparable frameworks.

In the explicit stated goals, for the TE at the four universities LIM is the university communicating a broader societal responsibility that their trained teachers should be able to respond to development needs of communities and enhance access for under-prepared and socially and economically disadvantaged students (pupils). The other three programmes focus on subject content knowledge and professional teacher skills (didactics and PCK).

Among the emphasised core competences, content knowledge is dominant at all four-teacher education programmes. GU highlights that TE can act as an instrument of social reform anchoring values such as equality, democracy,

*Table 3.3* Goals and Learning Outcomes

| *Gothenburg University (GU)* | *University of Limpopo (LIM)* | *Karlstad University (KAU)* | *University of the Witwatersrand (WIT)* |
|---|---|---|---|
| The goals include that students should have knowledge about:<br>- the academic subject and school subject<br>- general didactics, subject didactics and including subject methodology<br>- scientific theory as well as research methods<br>- knowledge of the relationship between scientific based knowledge and proven experience and its significance for the practice<br>- knowledge of children and young people's development, learning and needs<br>- knowledge of the organisation of the school system, relevant governing documents, curriculum theory and different pedagogical-didactic perspectives<br>- knowledge about assessment and grading | The goals of the programme is to:<br>- respond to the development needs of communities<br>- improving the university's overall academic standing and status<br>- enhance access for under-prepared and for socially and economically disadvantaged students<br>- produce suitably qualified and equipped career-oriented graduates<br>- provide opportunities for staff development<br>- diversify the university's funding streams | The goals include that students should have knowledge and understanding about:<br>- conceptual knowledge about the subject, subjects didactics, pedagogical knowledge, leadership<br>- traditional teacher knowledge base, general didactics, subject didactics including, e.g., assessment | Goals focus on holistic teacher development:<br>- subject matter competence<br>- pedagogical competence<br>- PCK |

*Table 3.4* Core Competences Emphasised

| *Gothenburg University (GU)* | *University of Limpopo (LIM)* | *Karlstad University (KAU)* | *University of the Witwatersrand (WIT)* |
|---|---|---|---|
| Content knowledge of school subjects<br>Research and experience-based TE and its significance for professional practice<br>Practical skills in teaching and everyday classroom work, the application of subject matter didactics and subject methodology, work independently and in together with colleagues<br>Teacher education as instruments of social reform which involves communicating and anchoring equality; democratic values, prevent and counteract discrimination and other abusive treatment; critical approach; ethical aspects such as human rights | Focus on competences related to PCK<br>- subject knowledge<br>- how learners learn<br>- curriculum knowledge<br>- classroom management<br>- assessment<br>and<br>- communication skills<br>- literacy, numeracy and Information Technology (IT) skills<br>- diversity<br>- work ethic<br>- critical reflection | Content competence, methodological competence and leadership competence<br>An academic professional aiming at developing teacher skilled at handling the profession's challenges in the daily school practice | Subject content and methodological specialisms<br>Professional learning skills<br>Transformation agency<br>Equality and human rights |

prevent discrimination, etc. Both LIM and KAU focus more on skills instrumental for teachers handling their daily work whereas WIT recognise TE as a possible transformation agency and promoting equality and human rights.

In the review of the TE programmes, there are some differences in what core values and competences foregrounded. The values and needs of Anthropocene may be taken for granted in the TE programmes of the north (human rights, equality and learning as transformation), while in the south, they are highly vibrant and urgent.

## Meta level discussion

By and large, the primary focus of the four-teacher education programmes is on subject content knowledge, skills and experiences needed for future careers as teachers. Across the four programmes there is an emphasis on: mastery of content, specialisation, theoretical knowledge, teaching methods/didactical competence, assessment skills and leadership competence, teachers' professionalism, communication skills, learning theories and school organisation. Thus, knowledge is constructed as an object – discrete to be mastered, transmitted and accumulated.

Drawing on Crossan et al. (2013), we argue that virtue education is essential to understand, feel, relate with sustainability and post-growth societies by altering this attitude of the human as a discrete occupier with discrete knowledge into nurturing a new way of life. Instead of perpetuating the knowledge accumulation approach with predefined practices the knower and knowledge is seen as co-evolving processes entangled with the wider world and changing as we ask deep and fundamental questions about ourselves and our imbrication with the natural world of which we are an integral and not necessarily unique part. However, as discussed earlier, of the virtues needed for TE in the Anthropocene, only a few can be traced in the steering documents compared in this chapter. In both of the Swedish TE programmes, knowledge is conveyed as a value approach thus beginning to open up the possibility of critique and transformation which is also recognised as important when engaging in socio-environmental issues. In case of South Africa, knowledge is viewed as a driving force for social transformation which in the light of the virtues needed for Anthropocene can be harnessed as a vital impetus towards a revolution of understanding of the human role in our world.

Regarding core competences, Gothenburg University in Sweden focuses on the importance of student teachers' and their education as instruments of social reforms which involves, e.g., democratic values, critical approach and ethical aspects, and all of which are crucial in the Anthropocene. The two-TE programmes in South Africa include critical reflection (in case of Limpopo) and transformation agency along with equality and human rights (the case of WITS). These competences are mainly targeted on the development of the individual learner. While this focus remains important, we argue for the need to go beyond the individual and teach in relational ways including connecting with local history and ecology.

To sum up, among the epistemological views, goals and core competences expressed in the steering documents for the four-teacher education programmes examined in this chapter a few of the sought virtues needed for teacher education in the Anthropocene can be found. It appears that dominant features of the four-teacher education programmes tend to have a utilitarian approach where knowledge accumulation and building functional knowledge are the focus of the education. We argue that TE should not be exclusively dominated by such practical education aimed at producing students who can successfully fit into society and school's daily context but rather underpinned by questioning of everything in its complexity.

## Conclusion

This chapter argues that a back to the roots shift needs to happen in education in order to have a chance to change our relationship with the environment to cultivate sustainable symbiotic relationships: asking the big and existential questions such as what are we here for, how big is space, how do we know what we know?

We argue that humanity has entered a new stage of evolution: epistemic evolution. Just as cultural evolution emerged against the background of biological evolution, epistemic evolution began as an aspect of cultural evolution and now dominates the global fate of humanity. It is characterised by the increasing dependence of global society on the achievements and further extension of science and technology in order to ensure its sustainability in the age of the Anthropocene. The historical development of knowledge is reviewed from an evolutionary perspective that introduces key concepts of a historical epistemology.

We emphasise that such an epistemic shift allows to begin to make the changes required to rebalance human interactions with the environment and sustain future existence, not as a slow change but an almost instant gear shift. Scientific knowledge needs to be recast as not a discrete complete set of tasks but as something ever evolving and being constructed and co-constructed with the knower who discovers themselves as an integral part of the world through asking questions.

It is thus time for TE to pose and address these existential questions pertaining to Anthropocene and prepare teacher trainees accordingly. As the popular chant of the 2018 school strike in Australia graphically argues, "There are no jobs on a dead planet!" This entails a change of perspective to seeing education as curiosity driven and expressing wonder at, and gratitude for nature in which we are fully embedded. This is at least as big a shift as the decolonisation of the curriculum whereby exclusively western ideas are questioned and interrogated alongside both indigenous ideas and ones from other cultures. Indeed, as some of the current malaise can be identified in a western colonial conception of the environment as a resource for profit, then its upending and transformation is both part of a decolonising education and ESD that both extend and deepen processes of decolonisation.

Of concern to humanity's stewardship of the planetary resources is an education system that recognises the interrelationship and synergies between the ecosystems that make up our planet. We need to ensure that our control of resources does not result in wanton destruction of the very legacies that would ensure the survival of the planet. We need to perceive and understand ourselves as an integral part of the ecosystem and not as a system apart. This dialogic ontology should inform our relationship with the environment.

By humans beginning to understand themselves as a crucial part of a wider environment, an environment they need to nurture and grow rather than a resource to be used and exploited, we can save ourselves from extinction. In that, education at large and TE in particular must emphasise curiosity, questioning and transformation.

## References

Adomßent, M., Fischer, D., Godemann, J., Herzig, C., Otte, I., Rieckmann, M., & Timm, J. (2014). Emerging areas in research on higher education for sustainable development-management education, sustainable consumption and perspectives from Central and Eastern Europe. *Journal of Cleaner Production*, *62*, 1–7.

Assadourian, E. (2017). EarthEd: Rethinking education on a changing planet. In E. Assadourian, & L. Mastny (Eds.), *EarthEd: Rethinking education on a changing planet* (pp. 3–20). Washington, DC: Island Press.

Barrett, M. J., & Sutter, G. C. (2006). A youth forum on sustainability meets the human factor: Challenging cultural narratives in schools and museums. *Canadian Journal of Science, Mathematics and Technology Education*, *6*(1), 9–23.

Battersby, M., & Bailin, S. (2011). *Critical inquiry*: Considering the context. *Argumentation*, *25*(2), 243–253.

Boylan, M. (2009). *Critical inquiry: The process of argument*. Boulder, CO: Westview Press.

Bray, J. N., Lee, J., Smith, L., & York's, L. (2000). *Collaborative inquiry in practice: Action, reflection and making meaning*. London: SAGE.

Brennan, M. (2017). Struggles for teacher education in the age of the Anthropocene. *Journal of Education (University of KwaZulu-Natal)*, (69), 43–66. Retrieved January 16, 2022, from http://www.scielo.org.za/scielo.php?script=sci_arttext&pid=S2520-98682017000200003&lng=en&tlng=en

Brennan, M. (2019). Changing teaching and teacher education in the 'Anthropocene'. *Journal of Research and Debate*, *2*(4), 1–6.

Chawla, L., & Cushing, D. F. (2007). Education for strategic environmental behavior. *Environmental Education Research*, *13*(4), 437–452.

Crossan, M., Mazutis, D., & Seijts, G. (2013). In search of virtue: The role of virtues, values and character strengths in ethical decision making. *Journal of Business Ethics*, *113*(4), 567–581.

Crutzen, P. J., & Stoermer, E. F. (2000). Global change newsletter. *The Anthropocene*, *41*, 17–18.

De Diego, A. M., & Díaz Carmona, E. M. (2018). Much more than knowledge: Virtue ethics and character education for degrowth. http://hdl.handle.net/11531/28638

Department of Higher Education and Training. (2015). National Qualifications Framework Act (67/2008): Revised policy on the minimum requirements for teacher education qualifications. *Government Gazette 596* (38487), 3–69.

Fadeeva, Z., & Mochizuki, Y. (2010). Higher education for today and tomorrow: University appraisal for diversity, innovation and change towards sustainable development. *Sustainability Science*, *5*(2), 249–256.

Gibson, K., Rose, D. B., & Fincher, R. (2015). *Manifesto for living in the Anthropocene.* Brooklyn: Punctum.

Guest, G., Namey, E. E., & Mitchell, M. L. (2012). *Collecting qualitative data: A field manual for applied research.* Los Angeles, CA: SAGE.

Hesselbarth, C., & Schaltegger, S. (2014). Educating change agents for sustainability-learnings from the first sustainability management Master of Business Administration. *Journal of Cleaner Production*, *62*(1), 24–36.

IPCC. (2021). https://www.ipcc.ch/report/ar6/wg1/#FullReport

Jickling, B., & Wals, A. E. (2008). Globalization and environmental education: Looking beyond sustainable development. *Journal of Curriculum Studies*, *40*(1), 1–21.

Johnson, B., & Villumsen, G. (2018). Environmental aspects of natural resource intensive development: The case of agriculture. *Innovation and Development*, *8*(1), 167–188.

Karlstad University. (2011). *Programme Study Plan Primary Education, Specialisation: Teaching in Pre-school Class and Grade 1–3 and Specislisation: Teaching in Grades 4–6.* Retrieved from https://www3.kau.se/utbildningsplaner/en/LAGLP_20132.pdf

Kimathi, F., & Rusznyak, L. (2018). Advancing professional teaching in South Africa: Lessons learnt from policy frameworks that have regulated teachers work. *Education as Change*, *23*(3), 1–25.

Kopnina, H. (2014). Revisiting education for sustainable development (ESD): Examining Anthropocentric bias through the transition of environmental education to ESD. *Sustainable Development*, *22*(2), 73–83.

Lafferty, W. M., & Eckerberg, K. (Eds.). (2013). *From the earth summit to local agenda 21: Working towards sustainable development.* London: Routledge.

Lave, J., & Wenger, E. (1991). *Situated learning: Legitimate peripheral participation.* Cambridge: Cambridge University Press.

Malone, K. (2017). Ecological posthumanist theorising: Grappling with child-dog bodies. In K. Malone, S. Truong, & T. Gray (Eds.), *Reimagining sustainability in precarious times* (pp. 161–172). Singapore: Springer. doi: 10.1007/978-981-10-2550-1_11

Orr, D. W. (1990). What is education for? In D. W. Orr (Ed.), *Hope is an imperative. The essential David Orr* (pp. 237–245). Washington, DC: Island Press.

Patrick, F. (2013). Neoliberalism, the knowledge economy, and the learner: Challenging the inevitability of the commodified self as an outcome of education. *ISRN Education.* Retrieved March 10, 2019, doi: 10.1155/2013/108705

Rieckmann, M. (2012). Future-oriented higher education: Which key competences should be fostered through university teaching and learning? *Futures*, *44*(2), 127–135.

Shulman, L. (1986). Those who understand: Knowledge growth in teaching. *Educational Researcher*, *15*(2), 4–14.

Silverman, D. (2011). *Interpreting qualitative data* (4th ed.). London: SAGE.

Sterling, S., & Jickling, B. (2017). *Post-sustainability and environmental education: Remaking education for the future.* Cham, Switzerland: Palgrave Macmillan.

Sund, L., & Öhman, J. (2014). On the need to repoliticise environmental and sustainability education: Rethinking the post political consensus. *Environmental Education Research*, *20*(5), 639–659.

Taylor, A. (2013). *Reconfiguring the natures of childhood.* London: Routledge.

Tilbury, D. (2011). Higher education for sustainability: A global overview of commitment and progress. *Higher Education in the World*, *4*(1), 18–28.

Swedish National Agency for Education. (2019). Curriculum for primary school, preschool class and the leisure center 2011. (Revised 2019). ttps://www.skolverket.se/publikationsserier/styrdokument/2019/laroplan-for-grundskolan-forskoleklassen-och-fritidshemmet-reviderad-2019

UNECE. (2012). *Learning for the future: Competences in education for sustainable development.* Geneva, Switzerland: United Nations Economic Commission for Europe (UNECE). Retreived 15th Feb 2022 from. https://unece.org/fileadmin/DAM/env/esd/ESD_Publications/Competences_Publication.pdf

United Nations Educational, Scientific and Cultural Organization (UNESCO). (2014). *Roadmap for implementing the global action programme on education for sustainable development.* Paris: UNESCO.

United Nations Educational, Scientific and Cultural Organisation (UNESCO). (2016). *Education for people and planet: Creating sustainable futures for all.* Paris: UNESCO.

University of Gothenburg. (2020). Programme Study Plans with specialization for Primary School and School Age Educare: *Revised 2020.* University of Gothenburg. https://www.gu.se/studera/hitta-utbildning/grundlararprogrammet

Wals, A. E., & Jickling, B. (2002). "Sustainability" in higher education: From doublethink and newspeak to critical thinking and meaningful learning. *International Journal of Sustainability in Higher Education*, *3*(3), 221–232.

Weldemariam, K. (2020). Reconfiguring environmental sustainability in early childhood education: A post-anthropocentric approach. *Environmental Education Research*, 26(12), doi: 10.1080/13504622.2020.1854690

Wiek, A., Withycombe, L., & Redman, C. L. (2011). Key competences in sustainability: A reference framework for academic program development. *Sustainability Science*, *6*(2), 203–218.

# 4 Linguistic opportunities and challenges of teaching science for sustainability in multilingual contexts

*Audrey Msimanga, Miranda Rocksén and Anna Maria Hipkiss*

## Introduction

In the light of post-humanist discussions on the Anthropocene, the shift from human focus to a rejection of the "separation between Nature and Human" (Latour, 2015, p. 146) has also brought new light on the role of language in the construction of what it is to be human (Pennycook, 2018). Questions of space, social practice and materiality are brought in, in order "to move away from the sedimented terminology of bilingualism and code-mixing towards a more flexible account of how people deploy different linguistic resources in their everyday practice" (Pennycook, 2018, p. 450). This chapter illustrates through examples of classroom interaction how language in the classroom is both a barrier and a resource/facilitator for teaching (and learning) about sustainability.

One of the most important functions of education is subjectification (Biesta, 2020). The subjectification concerns being able and having the freedom to act in different situations of one's life. It is a freedom which is integral to our existence. In the perspective of the present, and the rapid and global social and environmental changes, this means a severe challenge for teachers in science and technology. Freedom to act in relation to major changes in climate, biodiversity and social inequalities means that learning about sustainability is not something else, but an opportunity to slow down and make education more reflective and relational (Biesta, 2020), and this requires the learner's engagement. Access to quality education equips individuals and ultimately communities with the knowledge, skills, understanding, values and orientations to care about their environment and humanity in general (cf. Pennycook & Makoni, 2020; Whitehouse, 2011). Literate communities are enabled to engage on issues such as the well-being of the environment, on gender issues, matters of general being in communities and living peacefully as participant global citizens. Thus, education can be a tool with which to inculcate in learners the ideals of living sustainably, living and thinking about the importance of their decisions and actions on the environment not just locally but globally and the impact not just for this generation

DOI: 10.4324/9781003190158-6

but for future generations. The understanding of energy and natural resources, properties of chemical compounds and their effects on biological organisms, species biodiversity and ecology, develops in situated encounters where scientific vocabulary, reasoning and argumentation is successively appropriated. Teaching for and about sustainability therefore, as well as any other teaching, requires the teacher to enable learner engagement through the different linguistic resources that are available. In the learning about sustainability, as in the learning of science in different cultural, social and environmental contexts, there is a potential in the learners' language (Tang & Danielsson, 2018).

According to Norris and Phillips (2003) scientific literacy is a two-layered endeavour. Scientific literacy is about reading and writing in science, what they term the fundamental sense of scientific literacy, as well as about being knowledgeable in science as an informed citizen, the derived sense of scientific literacy. The process is facilitated through an education that includes knowledge about the natural sciences, and guiding principles within the subdisciplines, a scientific literacy for all. Yore (2012) presents a "vision" for science literacy as a: "specific illustration of disciplinary literacy in which the literacy component is recognized and valued for its functional roles in constructing understanding, persuading others of the veracity of these ideas, and reporting knowledge claims" (Yore, 2012, p. 2). This role of language is even more critical in contexts where science is taught and learned in a second or additional language as is the case in multilingual contexts (cf. Fang & Schleppegrell, 2008; Gibbons, 2006). As it is embedded in different communities' and funds of knowledge and discourses, Moje et al. (2004) presents the *skilled content literacy practice* – an alternative to binary conceptualisations of scientific literacy such as in- and out-of-school, everyday life and disciplinary, as well as a view on how scientific literacy may be developed in teaching and learning. Among the communities embedding the practice are families, popular culture, schools, peer groups, communities and social media (Information technologies). The skilled content literacy practice is constituted by three parts: *literacy skills*, that is encoding, decoding and comprehension; *knowledge*, that is concepts and word meanings in different contexts; and *discursive skills*, that is ways of making, using and communicating knowledge (Moje et al., 2004). By strategic design teachers may enable youths/students to draw on different funds of knowledge and resources and facilitate content and literacy development in a communicative third space.

The role of language in scientific literacy development and the role of language in science teaching and learning has gained prominence in local and global research. This is due to both globalisation with its related increased human mobility, in some areas creating "linguistic super diversity" (Duarte & Gogolin, 2013), and a renewed interest in interrogation of the political role of language in former colonies (Shizha, 2012). According to Shizha the process of decolonisation of science and science education can "place science into a multicultural space that respects cultural perspectives on scientific knowledge" (Shizha, 2012, p. 170). Such research has revealed the complexity of language issues in education and specifically in science teaching and learning, particularly in multilingual contexts/classrooms (see Msimanga & Erduran, 2018; Probyn, 2015).

In South Africa, for instance, literature records predominant stakeholder preference for English as the language of instruction mainly for its perceived role as the language of access, socially and economically. This even where English is not the home language or mother tongue of the teacher or the learner (Mthiyane, 2016; Rollnick, 2000). Decolonising education, as well as science, entails bringing in new perspectives, very much in line with Pennycook's ideas on language in post-humanist thinking, and bringing forth perspectives (and languages) that up until now have been underrepresented. Pennycook and Makoni (2020) suggest that a shift is needed from viewing language as "cognitive, literate systems rather than embodied and embedded cultural processes" (p. 127). They fear that removing human relations from language might lead to "the loss of shared forms of culture and knowledge" (p. 127).

Hence, it is essential for student teachers in science to develop knowledge about the role of language in the classroom, as both a barrier and a resource. In this context, and in line with United Nations' sustainable development goal number 4, about a quality education for all, we approach the social pillar of education for sustainable development in science teaching, through the concepts of scientific literacy and third space.

## Purpose

The purpose of this chapter is to investigate some of the linguistic resources that student teachers in our two countries could draw on as resources for teaching and learning, in order to work towards scientific literacy and sustainable development.

## Theoretical background

### *Scientific literacy and third space*

In developing a didactic model for science teaching based in language, Knain (2015) relates to the two visions of fundamental and derived senses of scientific literacy (Roberts 2007, 2011). By introducing the two visions of scientific literacy, the model captures science both within and outside the school as an institution. In the model, Knain represents disciplinary discourse as a secondary discourse in society, and separately, as a secondary discourse in the context of school. While school discourse consists of student-centred discourses and the disciplinary discourse of school science, the two are overlapping, constructing a *third space*. It is in this third space that teachers and students can meet in communication, where negotiations can take place about identity and the meaning of the subject content. Having access to this space is essential for science teachers in order to be able to support and develop students learning towards scientific literacy. Student teachers then need to learn about language use as both a flexible meaning-making resource and about the challenges involved in language use, when manifesting the "transformation and transition between students' life worlds and specialized practices such as science" (Knain, 2015, p. 154).

We are in this chapter adhering to a theoretical model for research on science literacy and language use presented by Wallace (2004). The framework consists of three intertwined components: (1) *authenticity* in science communication; (2) teaching about and providing opportunities for negotiation among *multiple discourses* in the science classroom; (3) Bhabha's (1994) notion of the creation of the dialogic *third space*. Following the words by Wallace (2004, pp. 905–910) authenticity means that in order to become scientifically literate, learners must appropriate academic scientific language into their own forms of communication. It represents a dimension of expression on the continuum between vernacular language and scientific language. The notion of multiple discourses implies that to become scientifically literate, learners need to recognise multiple discourse genres and learn to negotiate among them. Multiple discourses represent the continuum between private language, for example a question one pupil asks another pupil, and public language, for example in a textbook. Third space implies that to become scientifically literate, learners must be willing to enter the third space and collaborate in the construction of scientific meaning. Moje et al. (2004, pp. 45–46) writes about three ways of viewing third space: firstly as a way to build bridges from knowledge and marginalised discourses; secondly, as a navigational space, which means as a way of crossing and succeeding in different discourse communities; thirdly, as a space of cultural, social and epistemological change. The third way of seeing third space enables conversations between knowledges and discourses of different spaces that challenge and reshape the academic content literacy practices as well as knowledge and discourses of youth's everyday life.

## Method

In this book, we compare South Africa and Sweden, two countries that represent different language situations. The South African education language policy context is shaped by the provisions of the constitution and the Language in Education Policy (Department of Education, 1997). The constitution of South Africa recognises 11 official languages besides sign language and according to the Language-in-education policy any of these national languages can be used as the language of instruction beyond Grade 4 (year 5 of schooling). Thus, in terms of policy provisions, the South African education system provides for multilingualism. However, the decision on the selection of the language of instruction is left to each individual school. All instruction happens in the learners' mother tongue (natural language) for the first five years of schooling (Grade R-4) after which the school language policy comes into effect. This is where literature records an almost universal preference for English as the language of instruction. In Sweden the main language is Swedish. Sweden also has minority languages; Yiddish, Romany Chib, Sami, Finnish and Meänkieli, and a Swedish sign language. The Swedish language law (SFS 2009:600) states that people with mother tongue have the right to develop and use their mother tongue alongside Swedish. Swedish is the official language of teaching in pre-school, compulsory school and upper secondary education. The exceptions are a small number of schools with particular international profiles that use English as the language of teaching. Still, many classrooms are multilingual. In the year 2019, 300,000 pupils, between 6 and 15 years old, were eligible for native language teaching. Approximately 30% of the pupils in Grades 1–9 studied Swedish as second language (Swedish National Agency for Education, n.d).

To illustrate the linguistic resources available in Sweden, we share some results from a survey from the first course in the teacher education programme for primary teachers and out of school teachers taken twice, spring term 2020 and 2021. As part of a lecture on multilingualism in Swedish schools, students were asked to answer questions about their own language usage. One of the first questions was if they thought themselves multilingual and 179 out of 247 students (72%) responded that they see themselves as bilingual. When asked about which languages they speak, all but two students from spring term 2021 responded that they spoke two or more languages. There were examples of students naming four languages (e.g. Swedish, English, Estonian and German, or Swedish, Somali, Arabic and English). Some listed five languages (e.g. French, Spanish, Romanian, English and Swedish). Another survey question related to languages spoken on a daily basis by people near them (as in relatives and friends), and here the survey showed that there is a multitude of languages available to Swedish student teachers such as English, Somali, Arabic, Persian, Turkish, Thai, Tigrinia, Spanish and Kurdish. This abundance of languages highlights two things. First, that the linguistic landscape at teacher education has changed in the last decades and better represents the multilingual society in Sweden. Second, the breadth of linguistic resources student teachers bring with them is a resource teacher education cannot afford not making use of, and therefore it is urgent to find new ways of understanding and put them to work in different courses.

We are using examples of science teaching from our two countries, South Africa and Sweden to illustrate authenticity and multiple discourses. The first example is from Sweden and illustrates public and private language (Wallace, 2004) used in teaching and learning activities in chemistry. The second example, also from Sweden, illustrates scientific and vernacular language (Wallace, 2004) in practical activities in chemistry. In this context practical activity denotes students' experiments, chemical investigations and laboratory tasks. The third example, from South Africa, illustrates vernacular, scientific as well as public and private language in the teaching of biodiversity.

## Findings

### *Moving between languages in chemistry teaching*

The presentation and discussion that follows in this first section is based on two lessons in chemistry in a secondary school classroom (age 15) where students and teachers are part of a Content and Langauge Integrated Learning strand (CLIL). This means that about half of the lessons are taught in English (for example natural sciences and social sciences) and the other half are taught in Swedish.

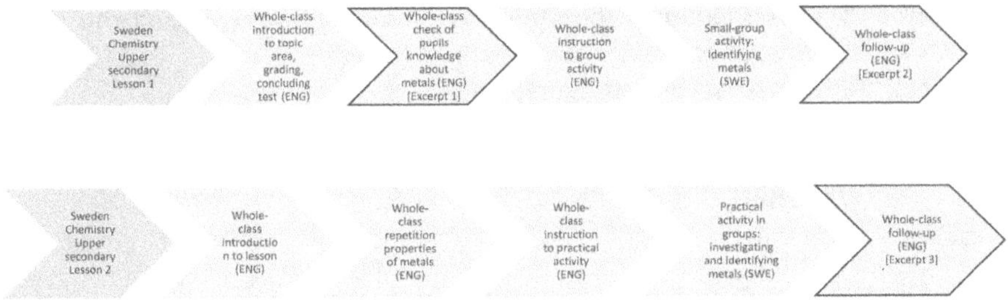


*Figure 4.1* Stages and prominent language use by the teacher in two analysed chemistry lessons in Sweden

For safety reasons, it was decided at the specific school that work in the laboratory ought to be conducted using the majority language, Swedish. From the CLIL setting follows that both students and teachers need to be accomplished English and Swedish speakers and, in the case of the students, they also need to develop disciplinary literacy in both languages. The two lessons consist of ten stages and the prominent verbal language use from the teacher's perspective is illustrated in Figure 4.1. Students used either Swedish or English and were free to choose even if there is an expectation to mainly use English.

Teacher-led activities are recognised by the teacher primarily speaking English, sometimes repeating himself in Swedish or translating individual words into Swedish. Our first excerpt illustrates the many semiotic resources involved in the check of pre-knowledge about metals. Present are the verbal languages of Swedish and English, a pull-down periodic table, deictic movements of teacher's pointing and students leaving their seats to interact with the poster in order to respond to a task:

*Excerpt 1* *Pre-knowledge check*

*(Translation in italics below original wordings in Swedish)*

Teacher: Okay then. What do you know about metals then?
Student 1: Det är hårt
*[It is hard]*
Teacher: Do you know anything about metals?
Student 2: There are many different kinds
Teacher: Different kinds and do you know how many of these [walks to the board and pulls down the periodic table poster] About one hundred elements that are metals? [words omitted] And about eighty-four is metals or are metals and [words omitted] and all are solids, they are solids in room temperature except this one [points with a sheet of paper]
Student 3: Vad är det?
*[What is that?]*

Teacher: Which is [some of the students stand up and move nearer the poster to see]
Students: Mercury [several students]
Teacher: Yes, Mercury, and in Swedish?
Student 4: Kvicksilver
*[Mercury]*
Teacher: Kvicksilver
*[Mercury]*
Student 5: Är den också giftig?
*[Is that poisonous too?]*
Teacher: Yes, extremely poisonous

Throughout the lesson, teacher-led activities are recognised by similar interaction including different semiotic resources such as movement between the two classroom languages and artefacts. The teacher bridges the scientific knowledge and the two languages in what can be understood as communication in a third space. In this lesson student and teacher meet in everyday understandings of the content, using Swedish. These meetings can further be compared to the "navigational space" (Moje et al., 2004) as the teacher provides opportunities for students to navigate between their everyday spoken understanding of the lesson content and thus preparing for being able to present their knowledge in other forms, as in written lab reports and examinations.

The second example is from the follow-up after a practical activity. Firstly, students were instructed in Swedish, to investigate if the compound (hematite) they were given was metallic using a magnet, then to add charcoal and again check if it was magnetic, and finally heat it up, let it cool and again check for magnetism. Secondly, the experiments were carried out in fume hoods and students spoke Swedish throughout. Thirdly, during follow-up, both Swedish and English were used in interaction together with the construction of a written reaction formula on the board:

*Excerpt 2 Follow-up after practical work*

*(Translation in italics below original wordings in Swedish)*

Teacher: så ni hade blodstensmalm hematit som inte ska vara magnetisk men ni fick den magnetisk efteråt
*[so you had bloodstone hematite which should not be magnetic but you got it magnetic afterwards]*
och vad var det som hände? What happened? [omitted] [writes on the board] okey, and you got iron, pure iron, because it was a metal but what did you also get?
*[and what happened?]*
Student: Koldioxid
*[carbon dioxide]*

Teacher: where did the oxygen go?
Student: syre
*[oxygen]*
Teacher: då kan man fundera varför tillsätter man kol. Vad har kolet för funktion?
*[makes you wonder why you add coal. What is the function of carbon/coal?]*
Student: det binder syret
*[it *binds* the oxygen]*
Teacher: yes, it combines with oxygen so you get carbon dioxide

In the excerpt the teacher uses the scientific terminology, hematite, as well as the historical name of the mineral, bloodstone, derived from the reddish colour. At this point the teacher states the observed magnetic properties because of the heating and focuses the chemical reaction when oxygen in hematite reacts with charcoal forming iron and carbon dioxide. Later in the follow-up the teacher talks about the blast furnace process which requires combustion at high temperatures and then mentions the historical deforestation of mining areas in Sweden. Again, this addition was in Swedish (translated below):

> Sweden has been, is perhaps still, world leading in this blast furnace process and iron ore. And, if you want to talk history, this is the reason to why large parts around Falun and Bergslagen, Dalarna were deforested. You need trees to be able to run these blast furnaces.

What can be concluded from the two examples is the many linguistic and other meaning-making resources at play, and the opportunities for meaning-making in the third space. English is the selected language of instruction and is used primarily by the teacher and some students. In order to make sure that subject content is understood for the lesson ahead the teacher also uses Swedish to focus certain concepts or actions and allows for students to use either language. The two examples focus introduction to a new work area and the first practical activity, which means the beginnings of the development of the disciplinary literacy to be tested at the end of the work area through a written test. What we learn from these examples, however, is how subject-specific vocabulary, subject-specific artefacts and the scientific explorations form a basis for the knowledge later to be expressed in writing and the opportunities for meaning-making that are created by the teacher when bringing in aspects from other contexts, be it his own background or historic knowledge.

### *Talking about chemical compounds*

The example in this section is from a secondary school in Sweden, using English as the language of instruction (year of preparation for the International Baccalaureate). In Sweden this particular education is provided in certain schools with an international profile. During this lesson pupils (age 15) are conducting a practical activity in chemistry. The students are learning about the properties of chemical compounds. While performing practical activities and experiments,

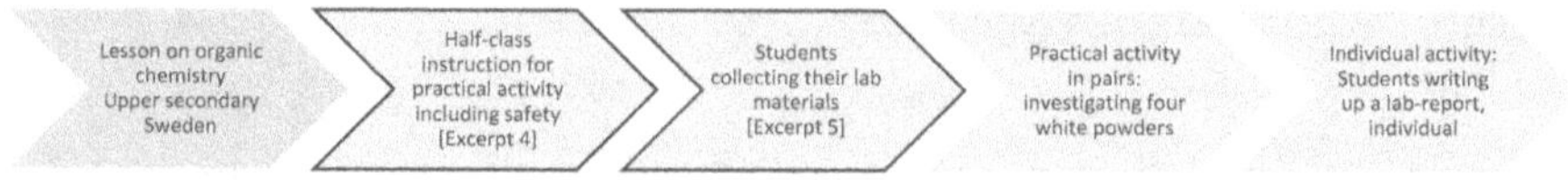


*Figure 4.2* Stages in a lesson about chemical compounds in organic chemistry in Sweden

they are also themselves exposed to the compounds, and any possible harms or effects from them on their own body. The example illustrates a classroom situation where the pupils are involved in learning basic chemistry, at the same time developing scientific literacy in the derived sense (Norris & Phillips, 2003) (Figure 4.2).

In the activity, pupils are supposed to make investigations on four white powders. Using different indicators, the task is to find the chemical compositions of the four compounds. The activity includes to observe and describe the properties of the powders and the chemical reactions involved. Before this short episode the teacher instructs the pupils, who are collecting the different materials to their benches. They work in pairs.

*Excerpt 3 "Is it dangerous?"*

Teacher: eh read through your instructions, see when I say ehm it's not really a proper hypothesis it's more what do you think the different things can be. A, B, C, D, just guess by looking at them without touching them and then you look at the ehm characteristics. You need to lose use lab aprons today because if you, for some reason, get this on your clothes it's it may stain

Student: Mhm

Teacher: So, same thing happens if you get it on your fingers, if you have stuff on your fingers it it's gonna stain, okay? [Questions?]

Student: How? [unhearable]

Teacher: It will give, it will leave brown marks on you, this is a brown solution

Student: If we have it on our things as well?

Teacher: Yeah so you see

Student: That's horrible

Teacher: If you get that on your fingers

Student: Yes

Teacher: You'll get brown spots on your fingers

Student: That's weird

Student: For the rest of your life?

Teacher: No, but it will, it takes a while for it to go away, it needs to be worn off. So, don't get it on your fingers. When you carry this bottle, carry it in the bottle, not in the top, cause it might not be securely fastened.

Student: So ehm, does it actually do anything else to us [as it is?]

Teacher: No. It's not dangerous.

Student: As long as you don't put it in your mouth.
Teacher: Don't put it in your mouth, no.
Student: Nothing dangerous happens?
Teacher: If you get too much iodine in your system you can be poisoned so don't drink it. Okay, I want you to wear lab coats and I want you to tie your hair back. Boys. I'm expecting braids now.

In line 1 the teacher instructs the pupils. The teacher repeats "A, B, C, D", which are the four samples of white powders. By saying "ehm it is not really a proper hypothesis", the teacher notes to the pupils that there are certain expectations on how to formulate a proper hypothesis, however this is not elaborated on. Instead, the pupils have to answer what compounds they think that these powders are. The teacher then tells them to use their lab aprons because one of the liquids they are using may stain. One of the students confirms this information saying "mhm". The teacher continues by saying that it also may stain their fingers. This starts an exchange in which the pupil articulates worries about the effects from this liquid on objects and fingers. The excerpt illustrates a situation in which pupils are asking questions about the compounds that they are supposed to work with. The teacher responds to the questions about safety and the possible harm on the body. The teacher also makes it clear that the activity is supposed to be performed using the specific genre of chemical investigations, which means for example reading the instructions, formulating a hypothesis and "black-boxing" the compounds naming them by letters A, B, C, D. In this way the situation in Excerpt 4 illustrates a meeting between the pupils' life world and the disciplinary language of school chemistry. In Excerpt 5, the pupils are standing around a desk by the white board where the teacher has placed the equipment.

*Excerpt 4* *"Which one is that?"*

*(Translation in italics below original wordings in Swedish)*

Teacher: so in the table it says physical characteristics so is it a powder is it a crystal is it when you touch does it feel slippery in your fingers is it does it have a color does it have a smell all of that would go towards the physical characteristics
Student 1: that's flour or
Student 2: or baking powder or flour
Student 3: Yeah
Teacher: try not to touch the one's using the same fingers, if you're touching one don't use the same finger when you put it in a different one because then you can contaminate your specimen
Student 3: You're holding C right
Student 4: mhm looks like brown sugar ha I don't even know [unhearable]
Student 1: sure brown sugar good guess which one is that?

Student 3: so that's this one right that's the eee
Student 1: Jag tror det var C CA [unhearable]
*[I think it was C CA]*
Teacher: you see the purple solution
Student 1: those are big isn't that [difficult to hear]
Student 1: no this feels weird
Student 1: the purple solution there
Student 1: Mhm
Teacher: if ehh
Student 1: you just [unhearable]
Teacher: protein is present in your sample then it will turn that color

While the pupils are collecting their material, the teacher provides some additional instructions, which has to do with firstly how to map the different characteristics, and a bit later pointing towards the risk of contaminating the samples. The pupils are looking at the samples and are guessing which powder they have got. One student then asks what sample another student is holding. A third student seems to hesitate and is unsure asking for confirmation. Then the first student suddenly turns to Swedish, saying "I think it was c" "[Jag tror det var c]". The teacher talks to a couple of other students, getting their attention about the purple solution, iodine, which will turn the colour of the sample if protein is present.

The short episode marks a practical activity in chemistry in a classroom in Sweden. It illustrates a classroom conversation about some of the fundamentals in chemistry education: to be able to investigate and systematically characterise the properties of different compounds, to develop understanding of chemical reactions, and how substances circulate in the environment and can have an effect on living organisms. In teacher education, these are situations that student teachers need to be prepared for. Not only by methodological practice, but also how to attend to and respond to pupils' worries about the effects of chemical compounds on their own body and the environment. This includes being knowledgeable in the disciplinary language of chemistry, but also being able to coordinate between the disciplinary language and the students' language, when they ask their genuine questions about how to understand the properties of chemical substances. Furthermore, this episode shows how Swedish appears as the supportive language, when two students are unsure about the labelling of two samples in a small chemical investigation.

### *Discussing owls in relation to biodiversity*

The excerpts in this section are taken from a whole class discussion on biodiversity in Grade 12 (age 18). The focus is on an activity exploring the role of owls as a legitimate part of biodiversity which should be conserved. In analysing this data, we try to do two things: highlight the teacher's intentions to create

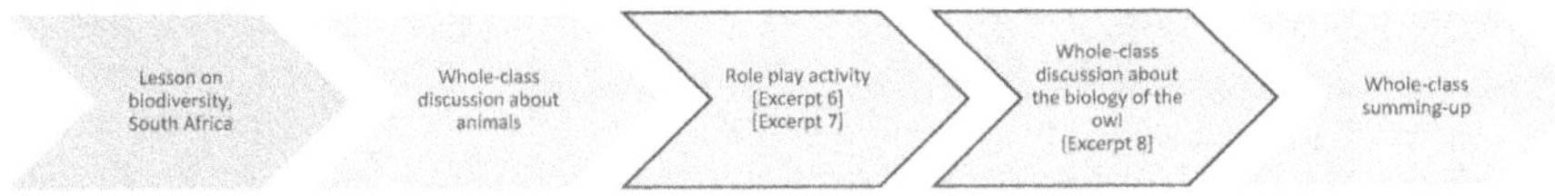


*Figure 4.3* Stages in the lesson about biodiversity in South Africa

dialogue by drawing on learners' indigenous knowledge and highlight cultural and language issues that emerge in the course of the discussion (Figure 4.3).

This was the introductory lesson in a series on biodiversity conservation. The aim was to explore learners' indigenous knowledge and beliefs or attitudes that might impact their understanding of the topic of biodiversity conservation. The focus of this lesson was the owl as an important link in the ecosystem and in biodiversity.

For many in Africa, owls are associated with evil and witchcraft. The owl is a sign of bad luck, it is bad to see or hear an owl call and it is taboo to touch it or its carcass (see, for example, Msimanga, 2000). The owl is, thus, not a welcome topic for discussion. However, the owl is part of the energy transfer process in the food chain and the food web as a predator. In the community where this lesson was presented there had been attempts by the local authorities to introduce owls as part of a programme to control rodents. The activity was designed based on this conflict between cultural views of the owl and its potential in bio-control of pests as well as biodiversity conservation in general. The lesson therefore addressed a very controversial topic for the South African multicultural context in which it was delivered.

The lesson started with a general discussion of animals and their role in the life of the learners. The teacher then steered the discussion to the following activity comprising a role play and concept cartoon responses (Keogh & Naylor, 1999). The concept cartoon represented four hypothetical learners, Ayanda, Bonelwa, Dumi and Chris (Figure 4.4).

**The role play**

***New notice from the City Council offices***

> *"In the interest of ensuring animal diversity and species conservation, government now requires all residents to keep at least one owl in their backyard"*

*As residents of this village, discuss the new notice. Remember everyone's idea is important, so*

- *Everyone must participate*
- *Everyone must say what they think*
- *You must listen to what others have to say*
- *You must always give reasons for what you say*

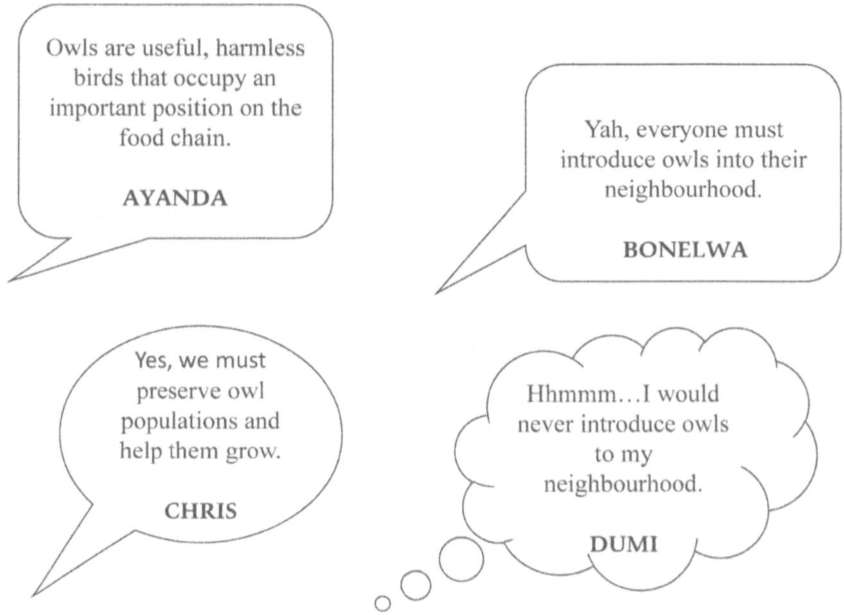


*Figure 4.4* Four responses to the question about the owl represented in a concept cartoon

In the first excerpt the teacher refers to Dumi, the fourth character in the concept cartoon and asks the class why they might agree with Dumi:

*Excerpt 5* *Why wouldn't you want an owl?*

*(Translation in italics below original wordings in isiZulu)*

Teacher: Dumi says "Hhhm no way I would never introduce owls to my neighborhood" (and then to the class) Now let's hear why wouldn't you introduce owls to your neighborhood? You should have reasons why you are against it. [student 1] why wouldn't you want to have an owl? I overheard you say "No no" Now tell us why.

Student 1: owl?

Teacher: yes what is it that you know? maybe it something that you read in a book in a newspaper. Or you saw in the television? Or if something that is er that you learnt from your parents or your grandparents or in your community. Remember different communities have different beliefs. So let's hear about that.

Student 1: ah Maam isikhova ah
*[no Ma'am an owl no]*

Class: [laughter]

Student 1: isikhova?
*[an owl?]*

Teacher: you keep saying "No no no" now tell us why
Student 1: isikhova?
*[an owl?]*
Teacher: no just be free. Oh someone wants to help him
Student 2: er Maam because many people believe that owls are used to bewitch other people
Class: yes yes

We refer to what happened in this episode as "learner cultural difficulty" in a lesson. Student 1's immediate reaction to introducing owls to his neighbourhood is "No, no, no". However, even after persistent probing, he could not articulate his reasons. Instead, he keeps asking the question "Owl?" and ultimately declares "No Maam an owl no". It seems that Student 1 was overwhelmed by the idea that anyone could even consider introducing owls into their neighbourhood and while it later became evident that he had an opinion or cultural understanding of owls he felt constrained to externalise his position. When Student 2 comes to Student 1's rescue articulating the belief by "many people" that owls are associated with witchcraft and by implication it would therefore be unacceptable for many to introduce owls to their neighbourhood. In subsequent turns the class almost unanimously agrees with Student 2. The cultural sensitivity of the topic had hindered engagement thus far and once Student 2 broke the silence there was no need to persuade her peers, the class was unanimous. This is significant for classroom discussions based on issues with strong cultural implications on two accounts. Firstly, the silence could be explained in terms of an established shared "understanding" of owls based on a common belief system, hence no need for argument. Secondly, the silence could be explained in terms of a need to be culturally appropriate. It would take a bold learner to challenge the common understanding at the risk of being viewed as taking sides with witches. If it is culturally accepted that owls are used in witchcraft, then one denies this at the risk of being viewed as supportive of witchcraft. Clearly, this has implications for teachers attempting to develop student scientific literacy in this context.

Later in the discussion we find the teacher attempting to steer the discussion to the original goal of the lesson, articulating and keeping explicit the teaching and learning agenda:

*Excerpt 6* *Continuation,*

*(Translation in italics below original wordings in isiZulu)*

Student 3: about an owl?
Teacher: Hmh
Student 3: Eh tjo Maam
*[oh no Maam]*
Class: [laughs]
Student 3: Tjo Maam li-violent igama lelo
*[Oh Maam the word is violent]*

Teacher: say it in vernacular
Student 3: No Maam it is a sign of death
Class: yes yes (general agreement)
Student 4: no Maam how?
Class: yes yes [talking noisily all at the same time]

At this point the teacher picked on Student 3 whose hand was not raised to speak but who had been talking to her peers on the side. Student 3's response is interesting for two reasons. In the first place she articulated her response in question form as if seeking clarification of the teacher's question when in fact she was expressing a reluctance to talk about owls. This is evident in her next two turns where she exclaims "oh no Maam the word is violent" "[Eh tjo Maam li-violent igama lelo]" referring to the word "death" which in some African cultures is taboo to even mention. Her response also seemed to change the tone of the conversation moving it from the seemingly more culturally accepted and "undisputed" association of owls with witchcraft to a more debatable belief about owls and death. While most of the class shouted agreement Student 4 raised a counter claim, "No Maam" phrased in the next two turns as the question "How?". Clearly in Student 4's culture (or in his personal conviction) the owl does not depict death thus he calls on Student 3 to provide evidence for her assertion. Once again, an instructional dilemma arises. While it is desirable for learners to challenge their peers' ideas and raise rebuttals, they themselves should be able to make their own claims (share their ideas as the teacher kept reminding the class) which Student 3 had declined to do earlier, according to the teacher's remarks at this point. What is interesting is the tension that emerges. On the one hand, the teacher attempts to balance the authoritative nature of the scientific language with the vernacular, to create dialogue. On the other hand, there is the pedagogical need to create a formal scientific discourse from the more informal engagement with socioscientific issues (e.g. Zeidler, Herman, Ruzek, Linder, & Lin, 2013). What is clear though is that the latter lends itself well to the use of learner languages as seen in this discussion.

In this next episode much later in the lesson, the discussion finally moves to the biology of the owl as Student 5 makes an unsolicited comment on the biology of the owl:

*Excerpt 7* *The biology of the owl*

*(Translation in italics below original wordings in isiZulu)*

Student 5: er Maam kahle kahle into nje okumele abantu bayazi ukuthi iskhova nje
*[what people should know is that an owl was not made to be]*
asidalelwanga ukuhamba emini sihamba ebusuku yikho nje abantu bethi iskhova siyathakatha
*[active during daytime it comes out at night that is why people say an owl is a witch]*

Class: Yes yes
Teacher Oh ok
Student 5: asikwazi kahle ukuhamba emini
*[it is not active during the day]*
Teacher: It is nocturnal not diurnal
Class: Yes
Teacher: in other words what he is saying is people are justified
Class: Yes
Teacher: to believe that it is used for a negative purpose because it is not seen during the day
Class: Yes yes
Teacher: and anything that is associated with darkness is usually er right believed to have implications of being used in witchcraft
Class: Yes
Teacher: That's what he is saying

The teacher uses Student 5's contribution to shift the focus of the discussion from beliefs about the owl, to the biology of the owl. Later, she follows up on Student 5's assertion that the owl is not active during the day by inserting the scientific terms *nocturnal* and *diurnal.* She then elaborates on Student 5's argument that in light of the evidence available people are justified to believe the way they do about owls. This way the teacher's intervention has potential to mediate engagement with the biology as well as emphasise the skills of argumentation/working with evidence.

Earlier in the lesson, the teacher had overlooked a learner's contribution based on the owl's biology, the fact that it fed on mice, but in the episode above and in the next one she took up the learners' ideas and began to use them to start working towards developing the scientific story of this and subsequent lessons on biodiversity:

*Excerpt 8* *The owl in the food chain*

*(Translation in italics below original wordings in isiZulu)*

Student 5: okunye Maam ukuthi iskhova sibalulekile for human beings
*[owls are very important for human being first]*
Iskhova okokuqala sidlaamagundwane lezezismoshela thina impahla
*[an owl eats rats that destroy our property]*
Teacher so that's a positive
Student 5: esinye sidli inyoka lezi ezisibulalayo
*[some eat snakes which kill us]*
Teacher: two that's a positive again
Student 5: so yikhoke ngithi sibalulekile iskhova
*[that is why I say an owl is important]*
Teacher: so it is very important in the food chain there will be less snakes there will be less mice in our environment

Student 5's argument in this excerpt further shifted the discussion to the importance of owls based on their biology. He, however, did not use the appropriate scientific terms, and this would keep the discussion within the realms of everyday knowledge. The teacher took up his idea and used it to point out both the positive role that owls play in the environment and the scientific explanation for this role, "so it is important in the food chain …". From this final example we conclude the strength shown in the pedagogical style of this specific teacher which by pursuing further this line of argument could have deepened learner engagement with the science content of this lesson even more. The teacher might probe learner understanding of the biology of the owl itself and that of the energy transfer in the food chain/food web. Also, the teacher might avoid providing the scientific terms herself and challenge her learners to do so. The key question here is: How best can teachers prepare to handle a lesson in such cultural and linguistic contexts? What can be learned for teacher education in this and similar examples?

## Discussion

Taking the starting point that one of the most important functions of education is subjectification (Biesta, 2020), science teacher education should expand on student teachers the freedom to act. Although we find the visionary concept scientific literacy theoretically useful, it is less useful for understanding what is at stake in actual science teaching and learning situations. In thinking about science and technology education and sustainable development, we find the notion of skilled content literacy practice interesting (see Moje et al., 2004). Teaching about sustainability could be strategically designed to draw on different sources of knowledges and discourses. This would be an alternative to binary conceptualisations of knowledge such as in- and out-of-school, everyday life and disciplinary, and what we believe a useful approach in the teaching about sustainability that also could be included and discussed in teacher education. In order to develop pupils' *literacy skills*, *knowledge* and *discursive skills* when making, using and communicating knowledge teaching would purposefully ask students to draw on families, popular culture, schools, peer groups, communities and social media.

There are many linguistic and semiotic resources at play in meaning-making. Teachers and students use the languages they can access to make their points, to make things clear, to understand. In this chapter, the examples show how teachers and students in both countries move between languages, between English, Swedish, isiZulu, on the continuum between public language, which is considered the disciplinary languages of chemistry and biology, and private language used between the students but also by the teacher when providing examples. In the teaching and learning the verbal activities are important, combining semiotic resources such as spoken language, textbooks, pictures, laboratory equipment, gestures and posters. Different funds of knowledges and discourses are drawn on. In the example from South Africa indigenous knowledge is used as resources in the teaching. In one of the examples from Sweden, the local mining history is used as a resource.

We acknowledge a third space of teacher and student interaction where student-centred discourses and the disciplinary discourse of school science overlap and where negotiations can take place about identity and the meaning of the subject content. We argue that it is a dynamic space in which science teachers must not only anticipate the forms of student engagement that are possible but also must respond in the moment to emerging student-centred and discipline-based discourses. Thus, student access and ability to engage in this space is often assumed. Our data illustrates the complexity of interaction in this space and the challenges facing teachers and teacher educators to support and develop student scientific literacy in context of language and cultural diversity. Teacher education programmes must be intentional in preparing student teachers for the forms of language use seen in the classrooms represented in this chapter; to be able to think about and work with language as a flexible resource; to leverage language to enable Knain's "transformation and transition between students' life worlds and specialized practices such as science" (Knain, 2015, p. 154). Future work could explore the extent to which teacher education programmes can respond to both the local, context specific teacher and student needs in the two countries as well as the broader common linguistic issues that have implications for scientific literacy development and educating for sustainability globally.

During the collaborative writing of this chapter as well as working together in the project has made us conscious about the many similarities science and technology teacher education is facing in our two countries. There are several ways this collaboration has had an impact on our construction of the text, our meaning-making and our understanding of science teaching in our two contexts. With the ongoing and complex process of decolonisation, which among many things entails understanding the Eurocentrism that has influenced both education and research (Pennycook & Makoni, 2020) we have learned how policy affects linguistic opportunities for school children. For example, the use of natural languages in schooling in South Africa has been a way to differentiate between those in power and those to be kept from power, increasing the distance between the scientific literacy of secondary school in English and that of local language primary classrooms. In Sweden, policy dictates that all languages in a classroom are to be seen as knowledge mediating resources in order to make access to higher education more attainable for all pupils. However, the reality is quite the opposite. Multilingualism is a global phenomenon present not only in classrooms in our two countries, but in every classroom. As a consequence, our focus has been on languages as resources for teaching and learning in science and technology. Furthermore, we argue that teacher education needs to prepare student teachers for the multilingual classroom (Duarte & Gogolin, 2013; Fang & Schleppegrell, 2008; Gibbons, 2006).

## Conclusion

We have in this chapter approached examples from science classrooms in which science teaching takes very concrete forms. The teaching and learning situations manifest a practice in which knowledge is used, the meanings of words are negotiated and available semiotic resources and forms of communication are put to work for different purposes. Science and technology teacher education must handle the challenges of the present, preparing student teachers to be able to use the full potential of the linguistic resources in their future classrooms.

## Acknowledgement

We are grateful to the students and teachers that throughout the years have contributed to the data included in this chapter and colleagues at our respective departments that made this possible.

## References

Bhabha, H. K. (1994). *The location of culture.* Abingdon: Routledge.

Biesta, G. (2020). Risking ourselves in education: Qualification, socialization, and subjectification revisited. *Educational Theory, 70*(1), 89–104. doi: 10.1111/edth.12411

Department of Education. (1997). *Language-in-education policy.* Pretoria: Department of Education.

Duarte, J., & Gogolin, I. (2013). *Linguistic superdiversity in urban areas: Research approaches* (Vol. 2). Amsterdam: John Benjamins Publishing.

Fang, Z., & Schleppegrell, M. J. (2008) *Reading in secondary content areas: A language-based pedagogy. Michigan Teacher Training.* Ann Arbor, MI: The University of Michigan Press.

Gibbons, P. (2006). *Bridging discourses in the ESL classroom: Students, teachers and researchers.* London: Continuum.

Keogh, B., & Naylor, S. (1999). Concept cartoons, teaching and learning in science: An evaluation. *International Journal of Science Education, 21*(4), 431–446.

Knain, E. (2015). *Scientific literacy for participation: A systemic functional approach to analysis of school science discourses.* Boston, MA: BRILL.

Latour, B. (2015). *Telling friends from foes in the time of the Anthropocene* (pp. 145–155). Abingdon: Routledge.

Moje, E. B., Ciechanowski, K. M., Kramer, K., Ellis, L., Carrillo, R., & Collazo, T. (2004). Working toward third space in content area literacy: An examination of everyday funds of knowledge and discourse. *Reading Research Quarterly, 39*(1), 38–70.

Msimanga, A. (2000). The role of birds in the culture of the Ndebele people of Zimbabwe. *Ostrich, 71*(1–2), 22–24.

Msimanga, A., & Erduran, S. (2018). Language, literacy and science learning for English language learners: Teacher meta talk vignettes from a South African science classroom. In *Global developments in literacy research for science education* (pp. 97–111). Cham: Springer.

Mthiyane, N. (2016). Pre-service teachers' beliefs and experiences surrounding the use of language in science classrooms: A South African Case Study. *Nordic Journal of African Studies, 25*(2), 111–129.

Norris, S. P., & Phillips, L. M. (2003). How literacy in its fundamental sense is central to scientific literacy. *Science Education*, *87*(2), 224–240. doi: 10.1002/sce.10066

Pennycook, A. (2018). Posthumanist applied linguistics. *Applied Linguistics*, *39*(4), 445–461.

Pennycook, A., & Makoni, S. (2020). Applied linguistics from a southern perspective. In A. Pennycook, & S. Makoni (Eds.), *Innovations and challenges in applied linguistics from the global south* (1st ed.). pp. 82–101. Abingdon: Routledge.

Probyn, M. (2015). Pedagogical translanguaging: Bridging discourses in South African science classrooms. *Language and Education*, *29*(3), 218–234.

Roberts, D. A. (2007). Scientific literacy/ science literacy. In S. K. Abell & N. G. Lederman (Eds.), *Handbook of research on science education* (pp. 729–779). New York, NY: Routledge.

Roberts, D. A. (2011). Competing visions of scientific literacy. The influence of a science curriculum policy image. In C. Linder, L. Östman, D. A. Roberts, P.-O. Wickman, G. Erickson & A. MacKinnon (Eds.), *Exploring the landscape of scientific literacy* (pp. 11–27). New York, NY: Routledge.

Rollnick, M. (2000). Current issues and perspectives on second language learning of science. *Studies in Science Education*, *35*(1), 93–121. doi: 10.1080/03057260008560156

SFS 2009:600. *Språklag [Swedish Language Act]*. Ministry of Culture. Retrieved 15th Feb 2022 http://www.riksdagen.se/sv/dokument-lagar/dokument/svensk-forfattningssamling/spraklag-2009600_sfs-2009-600

Shizha, E. (2012). Are we there yet? Theorizing a decolonizing science education for development in Africa. In A. A. Abdi (Ed.), *Decolonizing philosophies of education* (pp. 163–176). New York, NY: Springer.

Swedish National Agency for Education. (n.d.). *Elever med undervisning i modersmål och svenska som andraspråk (SVA) läsåren 2014/15-2019/20* [Data set]. Retrieved April 19, 2021, from https://www.skolverket.se/skolutveckling/statistik/sok-statistik-om-forskola-skola-och-vuxenutbildning?sok=SokC&verkform=Grundskolan&omrade=Skolor%20och%20elever&lasar=2019/20&run=1

Tang, K.-S., & Danielsson, K. (Eds.). (2018). *Global developments in literacy research for science education*. Cham: Springer International Publishing.

Wallace, C. S. (2004). Framing new research in science literacy and language use: Authenticity, multiple discourses, and the "Third Space". *Science Education*, *88*(6), 901–914. doi: 10.1002/sce.20024

Whitehouse, H. (2011). Talking up country: Language, natureculture and interculture in Australian environmental education research. *Australian Journal of Environmental Education*, *27*(1), 56–67.

Yore, L. D. (2012). Science literacy for all: More than a slogan, logo, or rally flag! In K. C. D. Tan, & M. Kim (Eds.), *Issues and challenges in science education research: Moving forward* (pp. 5–23). Netherlands: Springer.

Zeidler, D. L., Herman, B. C., Ruzek, M., Linder, A., & Lin, S.-S. (2013). Cross-cultural epistemological orientations to socioscientific issues. *Journal of Research in Science Teaching*, *50*(3), 251–283. doi: 10.1002/tea.21077

# Part II

# Experiences and Pedagogical Practices in Science and Technology Teacher Education

# 5 Application of the design process in the light of the Anthropocene

## A conceptual study

*Malose Kola, Maria Svensson and Anne-Marie von Otter*

## Introduction

In expressing his view about the revolution of the earth, Immanuel Kant (1996) predicted that not only would mother nature become extinct, but the very existence of thought would be extinguished leaving the cosmos in the cold conditions, barren and unable to reflect upon itself. Whatever has been created will remain wasteland, gratuitous and lacking the purpose. According to William et al. (2015), in their history, humankind (Homo sapiens) and their lineages had limited skills to influence the environment which they found themselves in. The ability to influence the environment began eventually during Palaeolithic as technology amplified and extended their skills from using stones as tools to lethality attainable through a pointed projectile. Technology became a progressively critical component in a system of prototypes (artefacts) leading to an increase in the likelihood of human survival and enabling population growth that emerged as a factor while discovering the new technologies. The link between technology and sustainability is evident in policy documents in South Africa and Sweden as well as in research and international reports (see, e.g., Department of Basic Education (DBE), 2011; Swedish National Agency for Education, 2018; UNESCO, 2009). All technology has an essential impact on society, nature and sustainability as a tool for building capacity for living and learning but also as a problem for sustaining the planet. Kamp (2006) emphasises that it is important to understand technology and sustainability from a global perspective, to have an overall picture. Education is the key to awareness of the possibilities and consequences of technology in a global context. Technology developed and used in one's own country affects people, communities and environments in many places on earth. As technology advances, the effect on the environment is increasing. For instance, there is environmental degradation, global warming, overpopulation and so on. An Anthropocene is a creation of human activities and of technology (Haff, 2014).

The question is to what extent these environmental issues are part of the design process where technology solutions are developed in technology education today. This is a key issue for technology teacher education as well as technology teachers in schools to prepare future generations, so that they, when

DOI: 10.4324/9781003190158-8

solving technology problems, understand the impact of technology on the environment. In this chapter we want to investigate two examples of how the design process is used in technology education in South Africa and Sweden in relation to the environmental situation in the Anthropocene. What is emphasised in the technology classroom in relation to sustainability? Conceptual analysis is used in order to understand the Anthropocene as a phenomenon. Furthermore, the curricula of both South Africa and Sweden were used as documents analysis. It was important to determine the extent in which both curricula promote sustainability or encourage learners to protect the environment. The authors examined the aspects of the curricula that put emphasis on the environment. Data was summarised and analysed narratively.

## Anthropocene

Climate change has led to intense discussions about the ability of modern human civilisation to affect the environment and opportunities to solve the problems created on the earth (Steffen et al., 2011). Subsequent to the discovery of the ozone layer cavity over Antarctica, which results from anthropogenic cause, the apprehension that the emission of large amounts of a colourless, fragrance-free gas like carbon dioxide ($CO_2$) could influence the energy balance at the earth's surface. Moreover, the anthropogenic effect has intensified the concern that human actions can harmfully affect a variety of ecosystem services that sustain human life including other species. Paul Crutzen, the Nobel laureate, avers that Anthropocene began in 1784, the era in which James Watt patented the steam engine, the symbolic of the beginning of the industrial revolution and the pollution of the atmosphere by burning coal excavated from the lithosphere (Bonneuil & Fressoz, 2017). This geological epoch, Anthropocene, is the effect of hundreds of years of human history. Oldfield et al. (2014) assert that through the Holocene and progressively since the transition to farming, humanity has successfully increased the inscription upon the earth. However, this is as a result of the era of industrial revolution which the inscription has progressed into a key strength influencing many global bio-geophysical cycles. Consequently, the industrial revolution became a vigorous, central and in some regards, a dominating dynamism in the earth system. By saying this, the relation between technology and the environmental changes is of great importance and is highly relevant in education today.

## Technosphere

Technosphere refers to the part of the environment where techno diversity extends its influence into the biosphere; the range or realm of human technological action; the technologically altered environment (Oxford Dictionary). Technosphere is a system where humanity is indispensable but at the same time subordinate to the changes that are taking place (Haff, 2014). In essence, the technosphere is autonomous. The property of technological autonomy instigates the platform for humanity to reflect about the deterioration of the environment from a human-centric to a system-centric point of view. However, this does not imply that humans are unable to influence its behaviour, but that the technosphere is rather inclined to resist endeavours to compromise its function. Haff (2012 & 2013) has previously argued that technosphere symbolises a new phase in geologic evolution of the earth. It is rather the global system whose operation strengthens the Anthropocene and consequently merits a particular consideration in our endeavour to grasp the role of humans in a nascent geological epoch.

To list but a few, the technosphere comprises across the globe the world's energy and resource extraction systems, power generation and transmission systems, communication, transportation, financial and other networks, governments as well as computers. There are six rules that humans apply which are considered as fundamental for the technosphere and these rules assist humanity to be informed about the dynamics of the Anthropocene (Haff, 2014).

Humans and the technosphere rules:

- The rule of inaccessibility – humans are elements of a large sphere which they did not comprehend, do not regulate and from which they cannot run away from.
- The rule of impotence – empowers humans to be exposed to the central phenomenon of the Anthropocene, the appearance of the large technological systems that have a habit to retaliate human intentions to or meddling with their functions.
- Leadership and the rule of control – an institution or nation do not survive without leadership.
- The rule of reciprocity – the physical constraint enforced on humans that they can address other systems such as automobiles or cell phones which are man-made products. The electronic products that are designed by humanity to a certain degree inspire the anthropocentric delusions that people developed and control far-reaching technology.
- The rule of performance – technology does not exclusively provide luxuries such as advanced utilities at home (microwaves, hoover or wine glasses) but essentials of life such as food and water which are accessible today as a result of the function of the technosphere.
- The rule of provision – the main system contributes to preservation of an appropriate environment for its parts in accordance with the rule of provision.

## The effect of technology

Technology is human-made hardware and knowledge applied to build objects to improve human abilities for executing activities they could not previously execute. The objects are either designed, manufactured and subsequently disbursed. Early on, technology was about simple implements such as axes, a jar of clay and parchment to expand her physical and mental abilities. Over time, more complex solutions have been adopted, such as rifles, binoculars and telegraphs.

In our time, the tools are often part of the socio-technical systems that have emerged and are now being used to expand our opportunities in terms of both quality and quantity. We can move over the earth at high speed with the help of cars, trains and airplanes, in a way that would not be feasible without the large socio-technical transport systems. Another example of a large socio-technical system is the communication systems that are built up by computers and cables, which give man access to an increased memory capacity, information and accessibility in a way that would otherwise be impossible (Kline, 1985/2003). To be successful in the latter a substantial system requires inputs such as raw materials, energy, labour and skills. Throughout the history of the human race, powerful capabilities were required by means of technology in order to change the manner in which they lived, established societies and in the process substantially affected the environment locally and worldwide.

## Fourth industrial revolution

The term fourth industrial revolution (4IR) collectively refers to a number of terms, which together describe the categorisation of a discipline (Lasi, Kemper, Feld, & Hoffmann, 2014). The 4IR is branded by the application of specific technologies such as big data, augmented reality, horizontal/vertical integration, simulation, cloud, additive manufacturing, cybersecurity, advanced robotics and internet of things (Petrilo, Cioffi, & De Felice, 2018). The impending 4IR signifies a far-reaching communication network that influences numerous institutions, suppliers, business, logistics, consumers, resources and so on. Every institution aspect boosts its structure in real time relying on the demands and the relevant sections in the network.

In order to harness the 4IR, the Department of Science and Innovation (DIS) in South Africa outline the following three set of megatrends that involve conjunctions of technology and disciplines which have a multiple system impact: (a) physical, (b) digital and (c) biological (Department of Science and Innovation (DSI), 2019). For example, digitalisation is anticipated to transform the work environment in ground-breaking methods, such as online jobs, crowdsourcing and the skill to labour-intensive workers and professionals to select when, where and how long they function. This might substitute the customary employment models. As a result, technology education ought to infuse this approach in the curriculum. Learners should be supported to embrace the 4IR and the nature of the activities provided should conform to new innovative means.

In Sweden the government 2017 presented a strategy for sustainable digitalisation in the 4IR (Näringsdepartementet, 2017). In this strategy the overall goal was to enable the use of digitalisation where people feel secure and become more and more competent to use and to develop innovations for the future. The five milestones presented in the strategy were: digital competence, digital security, digital innovation, digital management and digital infrastructure. These milestones are today implemented in the industry as well as in the education systems on different levels.

The 4IR have had an influence on the design process and Carvalho, Chaim, Cazarini and Gerolamo (2018) outline the following six main design principles and sustainability of 4IR with their most substantial influences to the environment:

- Interoperability
- Decentralisation
- Virtualisation
- Real-time capability
- Modularity
- Service orientation

*Interoperability* refers to elongated life cycle of the machine, reduction in industrial waste, quicker adaptation towards more effective methods. The concept "interoperability" does not have an explicit definition that is collectively shared by the overall community, it is prominently used to describe the core challenges in various systems and application scenarios (Pagano, Candela, & Castelli, 2013). *Decentralisation* refers to enhanced usage of local resources, improved application of accessible assets. *Virtualisation* is the reproduction of a device or its assets in virtual arrangement (Zahoor & Mir, 2018). Diverse processes from various applications operate at the same time, in so doing improving the efficiency and lessening the maintenance overhead. Virtualisation endeavours to minimise the complication by separating primary hardware and software. *Real-time capacity* denotes the improved adaptation to demand curves, improved application of resources, efficient reaction to energy supply modifications. *Modularity* refers to better utilisation of industrial resources, extended life cycle of the machine. The term modularity is broadly referring to the extent at which a system is composed of detachable and combined mechanisms (Lowell & Pollack, 2017). Broadly, the description of modularity in the context of system replicates a prominence on decomposing a system into compound disconnected sets of modules with explicit and limited mechanisms for communication or interface amongst those dissimilar sets. *Service orientation* refers to better-quality usage of ultimate products, better recycling and reuse prospects. Wang, Ali, Ramos and Vidgen (2013) assert that the advancement of IT that is based on service oriented computing put forward changes to old-fashioned software engineering. Service orientation advocates open, dispersed, vibrant and flexible systems that deliver commercial solutions.

## Education for sustainable development and technology

In relation to our interests in teacher's integration of sustainable development when teaching technology, teachers' own interpretations, capabilities and beliefs about sustainable development and the way it could be addressed through technology education are an important aspect to take into account (Pavlova, 2018). Earlier studies (see, e.g., Elshof, 2005; Pavlova, 2006; Pitt & Lubben, 2009) of teachers' perceptions of sustainable development in technology education show that, what teachers consider important to teach, is to a high degree reflected in their classrooms. The readiness among technology teachers to teach sustainable development in technology classrooms also depends on and reflects the specific contexts that occur in each country (see studies about Canada, the UK and Russia in Elshof, 2005; Pavlova, 2006; Pitt & Lubben, 2009).

Another important aspect to take into consideration is that there seems to be a technology-optimism among technology teachers concerning sustainability issues in relation to technology (Elshof, 2009; Knutsson, 2018). Sustainable issues "...are typically reduced to technical matters of eco-innovation: superficial 'tinkering practices' and tacit hopes that environmental problems can somehow be subject to 'techno-fix'" (Elshof, 2009, p. 135). Knutsson (2018) and Pavlova (2018) emphasise the need to include a more critical approach towards technology and sustainable development in technology education.

Research concerning technology education and sustainable development exists, but is not among the mainstream topics of research agendas (Pavlova, 2013). Studies of technology education and sustainable development are mainly discussed and focused on ecological design of products and the environmental impact or sustainability of products (Elshof, 2003; Stables, 2009). The social and cultural aspect of sustainable development is insufficient in technology education research. If economic aspects of sustainable development occur in relation to technology education, it is described mainly as something negative and in relation to developed countries (Elshof, 2003). Filho, Manolas and Pace (2009) investigated initiatives in schools and universities that exemplified existing practices where technology education and sustainable development have been integrated. They emphasise two inter-related characteristics of the subject technology, "(a) that it is not just a know-how subject, but a know-why subject [...] and (b) that it provides students with the opportunity to resolve problems and hence extend human capabilities" (p. 161). This implies that technology education must become more student-centred and include a more interdisciplinary approach to integrate various issues, including sustainability.

## Cradle-to-cradle

Cradle-to-cradle is one of the initiatives meant to sustain the environment. The concept cradle-to-cradle supports sustainable development. It is also regarded as a system of thinking which is based on the assumption that human endeavours can imitate nature's graceful system of safe and renewing production, by changing industries to sustainable inventiveness and eradicating the practice of waste. Natural ecosystems are founded on principles that could be modified by engineers in the industry (El-Haggar, 2007). McDonough, Braungart, Anastas and Zimmerman (2003) sketched 12 principles of cradle-to-cradle which engineers and designers should take into consideration in order to sustain the environment. Of these 12 principles, we here focus on four that are of high relevance in relation to technology education, namely: (a) designers need to attempt to make sure that all material and energy inputs and outputs are as naturally harmless as possible, (b) entrenched entropy and complication should be considered as investment for sustainability, (c) the design goal should be based on durability instead of immortality and (d) design of products, processes and systems should incorporate combination and interconnectivity with accessible energy and material flow.

Cradle-to-cradle is about seeing garbage as an eternal resource and doing the right thing from the beginning (McDonough et al., 2003). It is about making community and product development function in the same way as a healthy ecological system where all resources are used effectively. Embracing cradle-to-cradle principles establishes a cyclical flow of materials instead of an irreversible concept of cradle-to-grave. The materials used in industry bear resemblance to the nutrients that flow cyclically in normal ecosystems and can be circulated in biological or technical means. In accordance with the cradle-to-cradle notion, products are designed to be built of materials which can be safely developed, applied, recovered and re-applied while still sustaining their high value all through their entire life cycle. In order for the cradle-to-cradle system to be sustainable, all materials in products need to be kept clean and should not be mixed. Alternatively, there needs to be a separation system in place that can be applied after the item is discarded. Cradle-to-cradle methodology builds on the concept that "waste = food", meaning that what is considered waste can become food in a new product cycle. In this way valuable used material can be constantly cycled in closed loops and altered for reuse just like other products. The application of cradle-to-cradle design involves a modification of engineering techniques so that systems with closed loops of material flow are created. Cradle-to-cradle requires products to be designed in such a way to ensure that all materials can be classified into one of two cyclical systems: Biological cycle, materials that naturally biodegrade and can be returned to the ecological system. Examples of such materials are natural fibres and bioplastics. Or the technical cycle, physical transformation (e.g. plastic remoulding) and chemical transformation (e.g. plastic depolymerisation). It should be noted that materials in the technical cycle are therefore free to take part in any product, as long as they maintain their value. This intervention undoubtedly sustains the environment from waste generation and negative effects (El-Haggar, 2007). However, manufacturing industries could also gain advantage from the constant accessibility of products built of high value material despite the fact that the product has reached its lifespan.

The role of educational institutions in teaching strategies and techniques for designing eco-effective products and manufacturing processes, such as the cradle-to-cradle, needs to be further investigated. However, in a study with engineering students Gerber, McKenna, Hirsch and Yarnoff (2010) found that the students developed an understanding for the principle of cradle-to-cradle although they had difficulty reconciling their new-found interest in cradle-to-cradle design principles with the competing demands of client wishes, time constraints and limited access to and knowledge of materials.

There are limitations with using the cradle-to-cradle framework when working with technological solutions in the design process. When comparing the cradle-to-cradle framework with other methods for environmental impact assessment such as life cycle analysis (LCA), there are differences that one need to be aware of. An LCA is one of the important and versatile methods for analysing the environmental impact of products and services. An LCA study describes the entire industrial system involved in the production, use and waste management of a product or service. This means that a product is followed from its "cradle" when raw material is extracted from natural resources, through production and use to its "grave", when it is discarded (Baumann & Tillman, 2003).

The aim with the cradle-to-cradle is to inspire the creation of products with a positive impact on humans and the environment and the aim with an LCA is to systematically quantify all environmental impacts from life cycles of products and systems (Bjørn & Hauschild, 2018). In contrast to the LCA method, the cradle-to-cradle framework hardly addresses the energy consumption of products. Cradle-to-cradle thinking holds that as long as we use current solar income, energy is not a problem (Bakker, Wever, Teoh, & De Clercq, 2010). Another conflict between cradle-to-cradle and LCA arising due to the former's focus on quality is that the quantities of materials are not considered, as long as they are designed as technical or biological nutrients. This means that cradle-to-cradle does not support the idea of dematerialisation, which may lead to relatively material-intensive products. This is not a problem per se, but if the energy to produce the products is not current solar income (as mentioned above) or if the infrastructure to ensure the recycling of the product is not in place, it may result in a relatively bad LCA performance (Bjørn & Hauschild, 2018). Even though there are differences between both the LCA and the cradle-to-cradle, they could advantageously be used in technology education to raise awareness of the impact of technology and collaboration with the environment.

Cradle-to-cradle as a framework also has a relation to the main design principles and sustainability of 4IR. There is a clear connection to *Interoperability* where the life cycle of machines and the reduction of waste are highlighted. *Modularity* has a connection to the framework where the focus is on better utilisation of industrial resources by extending the life cycle of the machine through knowledge and increased efficiency in the mechanisms between different subsystems in complex production systems. In service orientation better-quality usage of ultimate products, better recycling and reuse prospects are the main focus which is in line with the two cycles, biological and technical, in the cradle-to-cradle framework.

## Design process

In technology education, problem solving involves finding solutions to people's needs by enhancing technological processes (Barak & Shachar, 2008).

Lewis (2006) adds that design comprises different procedural tactics whereby learners are provided with open-ended problems that compel them to develop alternative solutions. Visser (2009) avers that design is a cognitive act that involves the solving of ill-structured problems. Jonassen (2000) affirms that ill-structured problems are considered ill-structured because: (a) they have indefinite problem features or unknown elements, despite the individual's degree of self-confidence, (b) they have various solutions or no solution in any way, (c) there are various measures to evaluate the solutions, doubt about conceptual and procedural knowledge that are appropriate to seek solutions and (d) they frequently necessitate learners to make inferences and articulate personal interpretations concerning the problem encountered.

Many technology teachers find it difficult to strike a balance between teaching the design process, supporting the students to follow the design process, presenting content knowledge and providing individual attention (William, 2008). McCade (1990) acknowledges that although the design process is arguably the mainstay for teaching and solving technological problems, there are no substantial instructional strategies on how to teach the process. Van Dooren, Boshuizen, Van Merriënboer, Asselbergs and Van Dorst (2014) also affirm that technology teachers lack the appropriate conceptual and procedural knowledge required to teach the design process accordingly. The purpose of this study is to investigate the extent to which teachers support learners to consider environmental factors as they determine technological solutions.

## Design process in South Africa

Technology was introduced in the South African school curriculum (Curriculum and Assessment Policy Statement, CAPS) to encourage learners to become future engineers, technicians and artisans, careers that are in demand in modern society. According to the Department of Basic Education (Department of Basic Education (DBE), 2011), the purpose of technology is to, amongst others, stimulate learners to be creative and to develop the critical thinking skills required by a technology-driven society. The scope of technology in South Africa is to provide learners with opportunities to learn to, *inter alia*, solve problems in a creative manner, use authentic contexts embedded in everyday life and to evaluate existing technological artefacts and processes. Students are required to demonstrate the way in which a specific artefact meets the needs of humans in today's society, solves problems and creates opportunities. Most importantly, the criteria for teaching and assessing the design features, *inter alia*, are originality and aesthetics; value for money; and fitness for purpose; and easy to manufacture (Department of Basic Education (DBE), 2011). In the South African context, the design process refers to solving problems by applying the following processes: investigate – design – make – evaluate – communicate, see Table 5.1. Furthermore, the design process is a ground-breaking cooperative approach to solving problems (Department of Basic Education (DBE), 2011).

*Table 5.1* Design Process in South Africa (Department of Basic Education (DBE), 2011)

| | |
|---|---|
| **Investigate** | Problem/need/want |
| | Context/impact |
| | Research/questionnaire/interview |
| | Materials/suitable tools/required skills |
| **Design** | Initial ideas |
| | Free-hand sketches |
| | Design brief with specifications and constraints |
| | Plan using systems diagram |
| | Trial modelling |
| | Budget |
| **Make** | Choose tools/method/materials/resources |
| | Draw formal plans |
| | Draw flow charts/sequence of manufacture |
| | Make prototype/model/final product (considering safe working procedures) |
| **Evaluate** | Evaluate severity/urgency of problem/need/want |
| | Analyse solution using a systems diagram |
| | Evaluate solution in terms of design brief, specifications and constraints |
| | Evaluate product/process/manufacturing method/safety |
| | Evaluate impact/bias/an indigenous solution |
| **Communicate** | Report |
| | Present |
| | Advertise/poster using artistic graphics |

## Design process in Sweden

In today's society, knowledge of and about technology is needed to understand everyday life. The subject of technology in the Swedish school curriculum is described as a subject, which aims to enable students to develop their technical knowledge and technical awareness so that they can orient themselves and act in a technology-intensive world. Students should also develop an interest in technology and their ability to take on technological challenges in a conscious and innovative way. They need knowledge to better understand today's complicated technical phenomenons and contexts as well as knowledge about how technology affects society and humans. In Sweden, the subject has been regarded as a compulsory subject in compulsory school since the mid-1980s. However, the curriculum did not explicitly define how technology differs from well-established subjects such as science (Fahrman, Gumaelius, & Norström, 2015). After 1994, technology was defined as a subject in Sweden and differed from science with a clearer interdisciplinary character and environmental issues were incorporated into the curriculum.

The revised curriculum (Swedish National Agency for Education, 2018) for compulsory school has a core content that is based on three pillars: technical solutions, working methods for technical solutions and the connection between technology, people and society where technology development, problem solving

*Table 5.2* Phases in the Development Process of Technological Solutions

| *Phases* | *Processes* |
|---|---|
| Identification of needs | Identify and analyse technological solutions based on their appropriateness and function<br>Identify problems and needs that can be solved by means of technology and work out proposals for solutions |
| Investigating | Assess consequences of different technological choices for the individual, society and the environment<br>Analyse the driving force of technological development and how technology has changed over time |
| Proposing solutions | Learners' own constructions where simple mechanisms are applied |
| Designing | Documentation in the form of simple sketches, pictures and both physical and digital models<br>Controlling learners' own constructions or other objects by means of programming |
| Testing | Technological solution should be evaluated to determine functionality |

and technical systems have a large part (Swedish National Agency for Education, 2018). In Sweden, the design process refers to what is called: Working methods for developing technical solutions and includes: (1) identification of problems and needs, (2) investigations, (3) proposals for solutions, (4) design and (5) testing and evaluation. In the national curriculum for technology, there is also a strong emphasis on the consequences of technology choices based on ecological, economic, ethical and social aspects linked to sustainable development (Swedish National Agency for Education, 2018).

In companies LCA is a common method used when developing new technological solutions to identify the environmental impacts of a product or process at each stage of its life cycle (Baumann & Tillman, 2003). This method is also used by technology teachers when integrating sustainable development in technology education (Svensson & von Otter, 2016). Education for sustainable development is one of the international educational goals that are included in the Swedish National curriculum (Swedish National Agency for Education, 2018). In all subjects in the Swedish curriculum, it is important to create learning conditions for sustainable development based on democratic grounds. Recycling thinking, consequence thinking and system thinking describe teachers' perceptions and their teaching about technology and sustainability. In Table 5.2 different parts of the curriculum in terms of the design process are summarised.

## The aspects of sustainability in both countries

The term sustainability was formulated as the "*Nachhaltigkeitsprinzp*" around 1713. However, the German forestry lecturer, Von Carlwitz, conceptualised the notion by asking students what could be sustained in forestry (Weirsum, 1995). The response from students provided interesting description, namely: (i) the preservation of a prevalent product or product combination, (ii) the conservation of production capacity, (iii) the preservation of whole ecosystem instead of some particular components of such ecosystems or (iv) the maintenance of human systems which are rather forestry-reliance. The summation of the description is as follows:

> the need to ensure an equitable distribution of forest management inputs and outputs, and the need to arrange for such external conditions that forest managers are able to sustain these management practices
>
> (Weirsum, 1995, p. 42)

Furthermore, Brundtland Report of 1987 described sustainable development as conforming to the requirements of the existing generations, without rendering the natural resources base impotent, in order to enable the next generations to use these resources effectively. Wilderer (2007) affirms that sustainability is accomplished when the necessities of the existing conditions are met without hampering the capacity of the forthcoming generations to meet their own necessities. Sustainability is customarily used to define an objective that is somewhat unquestionably predictable (Weirsum, 1995). Kuhlman and Farrington (2010) affirm that sustainability is defined as conserving the welfare of the specific commodities for an unspecified period.

Guiding principles of sustainable development planning:

- Augment efficiency and expediency;
- Fortify the capacity of the upcoming generation to deal with the imminent challenges;
- Appreciate the bearing capability of the ecology which humanity if part of; and
- Reflect on the innate cultural values as treasure to be cherished (Wilderer, 2007).

In Sweden, *inter alia*, learners were expected to assess the consequences of different technological choices for the individual, society and the environment. Furthermore, amongst the core content of technology, learners are encouraged to adhere to safety when using technology, for instance when transferring information in a digital environment. Most importantly, learners are also encouraged to recycle and reuse materials in different manufacturing processes. Interaction between humanity and technology, as well as opportunities to develop technical solutions that contribute to sustainable development. Amongst others, the goals of the school in Sweden are that each learner should demonstrate respect and care for both the immediate environment, as well as the environment from a broader perspective.

Technology as a subject in Sweden is found in preschool to high school. Three different areas describe the content of the technology subject in Sweden: (1) technical solutions, (2) working methods for technical solutions and (3) technology, human and society. The national curriculum is not prescriptive, there is a lot of freedom for the teacher to choose different focus and approaches to teaching within the three content areas. Nevertheless, there are some focus and approaches that recur. Technology in the preschool focuses on letting children investigate artefacts and technical systems to make them aware of parts and wholes, to use different materials and tools to solve simple technical problems. In elementary school (children 6–12 years old) the design process is introduced and environmental aspects are connected to the process such as using recyclables when constructing and the consequences of technology for man, society and nature. When the students reach Grade 7, age 13 years old, teaching about the design process or LCA is often about the manufacturing process. The teaching of technology regarding the design process normally begins by learning the separate parts and what they stand for through material learning. When they are 14 years old they may choose a product they have in everyday life and examine how that product is made and what it is made of, such as a guitar or a toaster. When they are 15 years old, the design process is developed further and the students start by finding out how the raw material is produced, for example cotton, where they will describe the cultivation and working conditions for those who pick the cotton. The next step is to investigate how the cotton is washed and how the water is then purified. It is important to describe how cotton is transported to the spinning mill and what energy is used. In a final report they describe the textile factory's production of fabric to finished sweaters and take into account the ecological, social and economic environment from a quality technology perspective.

The DSI in South Africa developed a White Paper on Science, Technology and Innovation in order to advance the prospects of sustainability (Department of Science and Innovation (DSI), 2019). Amongst others, the objectives of this department are to enhance the enabling environment for innovation. South Africa has endorsed the sustainable development goals (SDGs) that were established by the United Nation General Assembly in the year 2015. In addition, South Africa is signatory to the Paris Agreement. These developments reflect the manner in which South Africa considers the aspects of sustainability and most importantly establishing a green economy. It is against this background that the Department of Basic Education (DBE) promotes sustainability. Hence, the definition of technology as a subject in South Africa is the use of knowledge, skills, values and resources to meet people's needs and wants by developing practical solutions to problems, taking social and environmental factors into consideration (Department of Basic Education (DBE), 2011, p. 8).

Amongst others, the aim of the South African curriculum (Department of Basic Education (DBE), 2011, p. 5) is to produce learners that are able to use science and technology effectively and critically showing responsibility towards the environment and health of others. Technology as a subject in South Africa is defined as "the use of knowledge, skills, values and resources to meet people's needs and wants by developing practical solutions to problems, taking social and environmental factors into consideration" (Department of Basic Education

(DBE), 2011, p. 8). Furthermore, the specific aims of technology are to support learners to develop technological literacy by providing them with the opportunities to, inter alia, appreciate the interaction between people's values and attitudes, technology, society and the environment. The unique features and scope of technology further give learners the opportunity to learn to, amongst others, deal with inclusivity, human rights, social and environmental issues in their tasks.

Technology education in South Africa is currently limited Grades 7–9. In Grade 7, learners are instructed to develop a working model of a hydraulic-syringe powered, linked lever reuse device using simple materials (Department of Basic Education (DBE), 2011). In addition, learners are provided with a case study on recycling scrap metals. In order to address the case study, learners are given an activity to tabulate a record of the waste produced by the school and these learners are instructed to develop a viable strategy to raise funds by recycling. In Grade 8, learners are exposed to the positive impact of technology and materials that are environmentally friendly and biodegradable. In addition, these learners are provided with the first case study to investigate the impact of plastic shopping bags on the environment. In responding to the case study, learners are instructed to sketch a report evaluating the effectiveness of using thicker, biodegradable plastic shopping bags which shoppers must buy. Furthermore, an additional case study is given to learners to investigate how waste paper and cardboard are recycled to produce new products for the packaging industry.

In Grade 9, when dealing with processing, learners are tasked to investigate the manner in which plastics are identified with codes and recycling. Learners are required to determine the properties of plastics (reduce-reuse-recycle). In addition, learners are provided with a case study which serves as an activity, to investigate manufacturing waste plastics into pellets for reuse. Lastly, learners are given another case study to investigate moulding recycled plastic pellets into products. When examining the nature of the activities that learners are exposed to, it is evident that the South African technology education curriculum considers the aspects of sustainability which has an effect on the Anthropocene. The nature of the activities confirms the definition of technology as a subject in South Africa: "the use of knowledge, skills, values and resources to meet people's needs and wants by developing practical solutions to problems, taking social and environmental factors into consideration" (Department of Basic Education (DBE), 2011, p. 8).

## Conclusion

Technology as a subject that is taught in the school provides an opportunity for learners to be exposed to the dynamics of the Anthropocene and how to mitigate the effect of technology in the environment. Due to survival or advancement humanity tempered with the environment. However, society has an opportunity to correct the impact of technology in the environment. As a result, the aspects of sustainability that have been introduced in both countries, South Africa and

Sweden, in the design technology (technology education) provide the basis for awareness and implementation. Learners are given activities that involve recycling and reuse. However, these learners are not aware of the effect of their actions. The culture of recycling and reuse is a springboard for sustainability, something that technology teachers should emphasise. Cradle-to-cradle is a new phenomenon that ought to be embraced including the means of energy renewal. These two concepts should be promoted in the technology classroom in most of the activities given to learners. The 4IR also outlined the features that advocate sustainability such as interoperability, modurability and service orientation. These concepts provide the basis for future research in technology design.

## References

Bakker, C. A., Wever, R., Teoh, C., & De Clercq, S. (2010). Designing cradle-to-cradle products: A reality check. *International Journal of Sustainable Engineering*, *3*(1), 2–8.

Barak, M., & Shachar, A. (2008). Projects in technology education and fostering learning: The potential and its realization. *Journal of Science Education and Technology*, *17*, 285–295.

Baumann, H., & Tillman, A.-M. (2003). *The hitch hiker's guide to LCA: An orientation in life cycle assessment methodology and application*. Lund: Studentlitteratur.

Bjørn, A., & Hauschild, M. Z. (2018). Cradle to cradle and LCA. In *Life cycle assessment* (pp. 605–631). Cham: Springer.

Bonneuil, C., & Fressoz, J. B. (2017). *The shock of the Anthropocene: The earth, history and US*. London: Verso.

Brundtland, G. H. (1987). What is sustainable development? *Our Common Future*, *8*(9). 1–40.

Carvalho, N., Chaim, O., Cazarini, E., & Gerolamo, M. (2018). Manufacturing in the fourth industrial revolution: A positive prospect in sustainable manufacturing. *Proceedia Manufacturing*, *21*, 671–678.

Department of Basic Education (DBE). (2011). *Curriculum and assessment policy statement (CAPS) of technology grades 7–9*. Pretoria: Department of Education.

Department of Science and Innovation (DSI). (2019). *White paper on science, technology and innovation: science, technology and innovation enabling and sustainable South Africa development in changing*. Pretoria: Government Gazette.

El-Haggar, S. M. (2007). *Sustainable industrial design and waste management: Cradle-to-cradle for sustainable development*. Burlington: Elsevier Academic Press.

Elshof, L. (2003). Technological education, interdisciplinarity, and the journey toward sustainable development: Nurturing new communities of practice. *Canadian Journal of Science Mathematics and Technology Education*, *3*, 165–184.

Elshof, L. (2005). Teacher's interpretation of sustainable development. *International Journal of Technology and Design Education*, *15*(2), 173–186.

Elshof, L. (2009). Toward sustainable practices in technology education. *International Journal of Technology and Design Education*, *19*(2), 133–147.

Fahrman, B., Gumaelius, L. B., & Norström, P. G. (2015, June). Technology education in primary school in Sweden: A study of teachers' views on teaching strategies and subject content. In *2015 ASEE Annual Conference & Exposition* (pp. 26–1497). ASEE Peer, Seattle, Washington.

Filho, W., Manolas, E., & Pace, P. (2009). ESD: Current discourses and practices and their relevance to technology education. *International Journal of Technology and Design Education, 19*(2), 149–165.

Gerber, E., McKenna, A., Hirsch, P., & Yarnoff, C. (2010). Learning to waste and wasting to learn? How to use cradle to cradle principles to improve the teaching of design. *International Journal of Engineering Education, 26*(2), 314.

Haff, P. (2012). Technology and human purpose: The problem of solids transport on the Earth's surface. *Earth System Dynamics, 3*, 417–431.

Haff, P. (2013). Technology as a geological phenomenon: Implications for human well-being. In C. N. Waters, J. Zalasiewicz, M. Williams et al. (Eds.), *A stratigraphical basis for the Anthropocene* (pp. 301–309). London: Geological Society.

Haff, P. (2014). Humans and technology in the Anthropocene: Six rules. *The Anthropocene Review, 1*(2), 126–136.

Jonassen, D. M. (2000). Towards a design theory of problem solving. *Educational Technology: Research and Development, 48*(4), 63–85.

Kamp, L. (2006). Engineering education in sustainable development at Delft University of Technology. *Journal of Cleaner Production, 14*(9), 928–931.

Kant, I. (1996). In M. J. Gregor (Ed.), *Groundwork for the metaphysics of moral* (pp. 37–108). Cambridge: Practical Philosophy.

Kline, S. J. (1985/2003). What is technology. In R. Scharff, & V. Dusek (Eds.), *Philosophy of technology. The technological condition* (pp. 210–212). Oxford: Blackwell Publishing.

Knutsson, B. (2018). Green machines? Destabilizing discourse in technology education for sustainable development. *Critical Education, 9*(3), 1–18.

Kuhlman, T., & Farrington, J. (2010). What is sustainability? *Sustainability, 2*, 3436–3448.

Lasi, H., Kemper, H. G., Feld, T., & Hoffmann, M. (2014). Industry 4.0. *Business & Information Systems Engineering, 4*(23), 239–242.

Lewis, T. (2006). Creativity: A framework for the design/problem solving discourse in technology education. *Journal of Technology Education, 17*(1), 35–53.

Lowell, J., & Pollack, J. (2017). The effects of environmental structure of the evolution of modularity in a patter classifier. *Proceedings of the ECAL 2017*, Lyon, France, 4–8 September.

McCade, J. (1990). Problem solving: Much more than just design. *Journal of Technology Education, 2*(1), 1–13.

McDonough, W., Braungart, M., Anastas, P. T., & Zimmerman, J. B. (2003). Applying the principles engineering of green to cradle to cradle. *Environmental Science and Technology, 37*(23):434–441.

Näringsdepartementet. (2017). För ett hållbart digitaliserat Sverige – en digitaliseringsstrategi. N2017/03643/D.

Oldfield, F., Barnosky, A. D., Dearing, J., Fischer-Kowalski, M., McNeil, J., Steffen, W. D., & Zalasiewicz, J. (2014). An Anthropocene review: Its significance, implications and the rationale for a new transdisciplinary journal. *The Anthropocene Review, 1*(1), 3–7.

Pagano, P., Candela, L., & Castelli, D. (2013). Data interoperability. *Data Science Journal, 12*, 19–25.

Pavlova, M. (2006). Technology education for sustainable futures. *Design and Technology Education: An International Journal, 23*, 733–475.

Pavlova, M. (2013). Teaching and learning for sustainable development: ESD research in technology education. *International Journal of Technology and Design Education, 23*(3), 733–748.

Pavlova, M. (2018). Sustainability as a transformative factor for teaching and learning in technology education. In M. J. de Vries (Ed.), *Handbook of technology education* (pp. 827–842). Switzerland: Springer.

Petrilo, A., Cioffi, R., & De Felice, F. (2018). *Digital transformation in smart manufacturing*. Rejeka: Intech Publications.

Pitt, J., & Lubben, F. (2009). The social agenda of ESD within design & technology: The case of the sustainable design award. *International Journal of Technology and Design Education, 19*(2), 167–186.

Stables, K. (2009). Educating for environmental sustainability and educating for creativity: Actively compatible or missed opportunities? *International Journal of Technology and Design Education, 19*(2), 199–219.

Steffen, W., Grienevald, J., Crutzen, P., & McNeill, J. (2011). The Anthropocene: Conceptual and historical perspectives. *Philosophy Transactions of the Royal Society, 369*, 842–867.

Svensson, M., & von Otter, A. M. (2016). Technology teachers' different ways of thinking about sustainable development in technology education. In 36th International Pupils' Attitudes towards Technology Conference (p. 33). Limerick, Ireland, 18–12 June.

Swedish National Agency for Education. (2018). *Curriculum for the compulsory school, preschool class, and school-age educare: Revised 2018*. Stockholm: Swedish National Agency for Education.

UNESCO. (2009). Information Society Policies. *Annual World Report 2009*. Intergovernmental Council. Paris: UNESCO.

Van Dooren, E., Boshuizen, E., Van Merriënboer, J., Asselbergs, T., & Van Dorst, M. (2014). Making explicit in design education: Generic elements in the design process. *International Journal of Technology and Design Education, 24*, 53–71.

Visser, W. (2009). Design: One, but in different forms. *Design Studies, 30*, 187–223.

Wang, X., Ali, W., Ramos, T., & Vidgen, R. (2013). *Agile and lean service-oriented development: Foundations, theory and practice*. Hershey: Information Science Reference.

Weirsum, K. F. (1995). 200 years of sustainability in forestry: Lessons from history. *Environment, 19*(3), 321–329.

Wilderer, P. A. (2007). Sustainable water resource management: The science behind the scene. *Sustain Science, 2*, 1–4.

William, P. J. (2008). Using DEPTH as a framework for determination of teacher professional development. *International Journal of Technology and Design Education, 18*, 275–284.

William, M., Zalasiewicz, J., Haff, P. K., Schägerl, C., Barnosky, A. D., & Ellis, E. C. (2015). The Anthropocene biosphere. *The Anthropocene Review, 2*(3), 196–219.

Zahoor, S., & Mir, R. N. (2021). Resource management in pervasive Internet of Things: *A survey. Journal of King Saud University-Computer and Information Sciences*, 33(8), 921–935.

# 6 Discussing pedagogical content knowledge and ESD for science teaching

## Bridging theory and practice in the Anthropocene

*Sally Windsor, Maria Svensson, Washington Dudu and Marlene Sjöberg*

### Introduction

A dilemma in the preparation of new teachers around the world is that of the theory-practice divide (often also referred to as a gap). Much literature (see, for example, Brandt, Bürgener, Barth, & Redman, 2019; Hume, Cooper, & Borowski, 2019; Shumba & Kampamba, 2013) has posed the question to those becoming teachers, and the profession at large, to consider what is most important when learning to teach, theoretical knowledge or practical knowledge? This theory practice gap "is a perennial issue in teacher education" (Gravett, de Beer, Odendaal-Kroon, & Merseth, 2017, p. 370) and has been described by Kessels and Korthagen (1996) as the difference "between our [i.e. teacher educators] words and the students' experiences that we cannot bridge" (p. 2). However, we start with the understanding that if this gap exists, and is problematic for new teachers, providing conversational opportunities to discuss theory after engaging in practical experiences can begin to close the gap. These types of conversation are important in teacher education programmes because they "allow student teachers to become conscious and thoughtful about their actions, as opposed to using trial and error to deal with confusing and problematic situations" (Sööt & Viskus, 2015, p. 1941). Confusing and problematic situations in the school classroom, we argue, are ubiquitous in the Anthropocene.

Teacher educators, in turn, must listen to what student teachers say about the gap and what professional and pedagogical knowledge they need to teach in the Anthropocene. Scholars of teacher education have long found that the provision of opportunities to bridge the theory practice divide is critical. And while the notion of critical reflection has been well researched for decades and agreed as vital to the process of teaching and professional development it is

> however, questionable whether teachers are adequately trained to follow up on this during their process of education or are provided with adequate conditions in their everyday pedagogical practice (school management's

DOI: 10.4324/9781003190158-9

> support, enough supervisors, time, etc.) and given support when reflecting on their teaching practice
>
> (Saric & Steh, 2017, p. 70)

To do our job as teacher educators responsibly, we need to come to grips with a complex reading of the world, including content knowledge and pedagogical knowledge for global citizens and members of multiple communities, including those of worldwide educators in the 21st century.

## Purpose of the chapter

This purpose of chapter is to contribute to a better understanding of what student teachers believe they need for Science, Technology, Engineering and Mathematics (STEM) and Education for Sustainable Development (ESD) teaching in the Anthropocene, by drawing on reflections of student teachers in conversation with their mentors and/or teacher educators. Through professional reflective conversations focused on in-school teaching experiences between student teachers and teacher educators in South Africa and Sweden, we are able to report on what student teachers say, in different contexts, that they need for teaching STEM in the Anthropocene.

### *Research questions?*

- In what way do professional conversations between student teachers and teacher educators allow for and promote reflection on teaching experiences?
- What new insights are gained between teacher educators/researchers from the north and south regarding new ways of developing and facilitating pedagogical content knowledge (PCK) and ESD in science teacher education?

In this chapter, we use small glimpses of data from two different studies to illustrate our position. The data explicitly shows us some things that student teachers highlight as needed for their teaching in the future. We compare how STEM teacher educators in selected contexts in South Africa and Sweden converse with STEM student teachers about how PCK and ESD can help teachers and students work towards the Sustainable Development Goals (SDGs) (United Nations (UN), 2015). Research (Sjöberg, 2020) looking at reflective conversations where student teachers completing STEM teacher education have focused conversations about practice with a teacher educator has shown these conversations to be beneficial to the development of PCK. Thus, in this chapter, we discuss PCK and ESD in the Anthropocene, compare PCK in STEM teacher education in the two different cultural and political settings in South Africa and Sweden and exemplify how student teachers' professional reflective conversations could be used to enhance the development of PCK in this time of the Anthropocene. We argue that opportunities for teacher educators and student teachers to discuss

PCK after teaching practice must be provided in initial teacher education (ITE) in both countries to a greater extent.

## Theory

Before briefly outlining the concepts of ESD and PCK in terms of STEM teacher education specifically it is important to understand what new dimension the Anthropocene brings to the questions of teacher preparation, and why it is increasingly important for teachers to be able to build and maintain bridges between theory and practice within it. As outlined throughout this book, the Anthropocene is a term describing a new geological epoch characterised by the situation where human enterprise has grown so dramatically that it has destabilised geological processes that the relatively stable Holocene epoch endured (Crutzen, 2002; Steffen, Crutzen, & McNeill, 2007). In other words, "the 'Anthropocene' is a concept around which action has been mobilised, as well as hard thinking" (Brennan, 2017, p. 44). With this understanding of the unstable and uncertain future the Anthropocene posits, how does teacher education renegotiate the types of preparation needed by future STEM teachers?

### *Education for sustainable development (ESD)*

Education is at the heart of efforts both to adapt and change and to transform the world within which we live (UNESCO, 2015, p. 3). ESD as a phenomenon is almost entirely shaped by the United Nations Education, Scientific and Cultural Organisation (UNESCO) in response to the United Nations (UN) notion of sustainable development and the current SDGs. The current UNESCO ESD programme is called ESD for 2030 and specifically "aims to build a more just and sustainable world through strengthening ESD and contributing to the achievement of the 17 Sustainable Development Goals" (UNESCO, 2019). As Evans (2019) confirms "it is clear that the scene for the fifteen years between 2015 and 2030 is set, the characters are fixed and education, specifically teachers, are (re)confirmed as having a key role in the achievement of sustainability" (p. 13).

### *Pedagogical content knowledge (PCK)*

PCK is a term defined by Shulman in 1987 to describe "the knowledge used to transform subject matter content into forms more comprehensible to students" (Shulman, 1987, p. 13). It differs from knowledge of the subject or content, and from general pedagogical (and in the continental European tradition – *Didaktik*[1]) knowledge in that it is the type of knowledge that is held by effective teachers who "know how students tend to understand (and *mis*-understand) their subjects" (Grossman, Schoenfeld, & Lee, 2007, p. 205). PCK is developed through engaging with both theory – learning about what to look for, and practice – where teachers can gain experience in identifying the patterns of learning in classrooms. It is well understood that professional knowledge, and in particular PCK for STEM teaching, develops over time and interplays with professional experiences in the field (see, for example, Appleton,2013; Park & Oliver, 2008; Sjöberg & Nyberg, 2020).

To meet the demands of developing PCK in a holistic way (both in theory and practice) and in relation to what we today know about the Anthropocene, we suggest an interactive and learner-centred model that highlights the importance of professional reflective conversations between student teachers, their mentor teachers and teacher educators (Kriewaldt, Nash, Windsor, Thornton, & Reid, 2018). The model builds on exploratory, action-oriented and transformative learning through developing student teachers' reflective capacities in and after these conversations (Windsor, Kriewaldt, Nash, Lilja, & Thornton, 2020). In this chapter, we discuss ESD and PCK, compare PCK in science and technology teacher education in the two different cultural and political settings in South Africa and Sweden and exemplify how student teachers' professional reflective conversations could be utilised in developing PCK.

### *Professional reflective conversations*

It is generally understood that when teachers have opportunities to discuss and reflect upon their teaching practice, their teaching practice is enhanced (Clarke, 1995; Eckerman Pitton, 2006; Timperley & Alton-Lee, 2008; van Kruiningen, 2013). Professional conversations between student teachers and more experienced teachers are crucial and are considered "central to developing student teachers' cognitions that underlie their professional knowledge and performance" (Timperley, 2001, pp. 111–112). Providing opportunities for student teachers and teacher educators (who generally hold reserves of knowledge about teaching practice) to engage in conversations may "bridge the gap between theoretical perspectives on teaching and learning and teaching practice" (Sjöberg & Nyberg, 2020, p. 227). As Loughran (2014) reminds us, it is not only important for student teachers to have teaching experiences, but it is critical to reflect on those experiences. Professional conversations provide those learning to teach with reflective opportunities that can foster teaching practices that can bridge the science and the art of teaching (Mun Ling & Marton, 2012).

## Method

As mentioned above this chapter is predominantly theoretical yet it draws upon empirical data from two separate studies in South Africa and Sweden. In both cases the data are excerpts from recorded professional conversations conducted by teacher educators with STEM student teachers about their experiences after teaching practice periods. Conversation transcripts from both contexts were revisited by the authors of this chapter and the transcript data reanalysed. This revisiting was done with the focus of "listening" to what the student teachers said explicitly and implicitly about what they need to develop PCK and their teaching regarding ESD. Revisiting the data which involved a "refram[ing of] methodological intentions" (Hipkiss, Windsor, & Sanders, 2020, p. 351) allowed the authors of this chapter to compare the discussions that took place between student teachers and teacher educators about PCK and ESD in the two contexts.

The Swedish data comes from a doctoral study that explored "student teachers' professional conversations upon teaching experiences, with pedagogical content knowledge, PCK, as an analytical lens" (Sjöberg & Nyberg, 2020, p. 226). The South African data is drawn from interviews conducted by teacher educators, who asked student teachers from a range of STEM areas about ESD specifically. One of these interviews was a group interview which provided a large and rich example of a conversation between the student teachers and their teacher educator.

To contextualise the findings the teacher education programmes in South Africa and Sweden as they relate to STEM teacher education are briefly outlined. Following that, the method of analysis is described.

### *Sweden*

At the University of Gothenburg, Sweden, there are a number of teacher education programmes and the first unit of study for all programmes focuses on pedagogical knowledge, *didaktik* and the history and development of the Swedish school system. This first course in teacher education programmes focuses on how learning perspectives and *didaktik* traditions interact in the meeting with different content, students and classroom situations. The aim of the first course in teacher education is, among other things, to introduce the basic questions of *didaktik* – *what* is being taught and learnt, *how* is this affected and *why* – with what purpose – is this carried out. In this course, the relationship between the scientific basis and proven experience (teaching practices) as well as the ability to take the learners perspective is explicitly discussed.

During four school practice courses (4 × 7.5 Higher Education Credits), the student teachers are expected to develop practical knowledge about the teaching practice through experience in school classrooms. By the end of these practical courses student teachers are expected to be able to plan and carry out teaching where all pupils are catered for and learning is based on well-motivated *didaktik* and pedagogical choices and with relevant subject knowledge in an independent manner.

Professional conversations are conducted during the four school practice courses when the students discuss their planned teaching with the mentor and during a seminar together with teacher educators and student colleagues during and after the school practice. The nature of such conversations is often reflective but not necessarily in a structured way with the aim to bridge the gap between theory and practice concerning a specific subject or content. During this teaching sequence the students discuss their plan with a science pedagogical specialist and a science teacher educator and reflect on their experiences of the lesson in pairs, and in a big group with student colleagues, teacher educators and science teachers.

The professional conversations that provided data for this chapter were recorded and transcribed for a doctoral study. In that study Sjöberg (2020) explored how student teachers and a team of science teachers engaged in professional conversations about science teaching. The phase of the study looking specifically at student teacher's reflective contributions to the professional conversations was analysed using PCK as a framework for content analysis.

### *South Africa*

In South Africa, the chapter's specific focus is upon student teachers enrolled at North West University. At North West University, there are a number of Bachelor and Masters education programmes for teacher education. In all programmes, while PCK is not a mandated (or core) component, it is assumed that it should be incorporated into all courses.

With the understanding that teaching is a profession the work-integrated learning (WIL) component commonly known as "teaching practice" (when student teachers go into schools to do their practicum) is important in South African teacher education programmes. In the South African context, student teachers go into schools twice a year where they are supervised by school-based mentors and university lecturers. At North West University student teachers undertake periods of teaching practice in school twice a year for four years. It is during this period that student teachers implement what they learn during methodology modules which are either stand alone or infused with content modules. Opportunities for conversations in the South African settings are centred on the teaching practice periods. Student teachers are able to discuss PCK concepts throughout these periods with their school-based mentors. Conversations are also conducted by teacher educators (university lecturers) but these are in the context of summative assessment rather than professional development. Some conversations occur during micro-teaching sessions (this is when teacher educators allow student teachers to teach each other during methodology modules) practising how to teach certain topics of the curriculum. The conversations reported in this chapter were conducted at North West University with student teachers in their second and third years situated in the North West province of South Africa. The university has three campuses, namely, Mafikeng, Potchefstroom and Vanderbijl Park and the conversations took place at the first mentioned campus situated in the north-west part of South Africa. The professional reflective conversations were conducted, recorded and transcribed for the purposes of this chapter. While discussing the findings in relation to PCK and ESD in the Swedish study, the authors wished to get a sense of the similarities and differences in the student teachers' reflections in the two settings.

## Content analysis

Content analysis is a research method for subjective interpretation of the content of text data through a systematic classification process including coding and identifying themes or/and patterns. We employed a conventional content analysis process as described by Hsieh and Shannon (2005) – an approach used when existing theory or research literature on a phenomenon is limited and that allows the categories/themes to emerge from the data in an inductive way. During this process, the focus was on the contextual and content-related meaning produced by the participants (Graneheim & Lundman, 2004; Hsieh & Shannon, 2005; Schreier, 2014).

In the first phase of the analysis the authors read the transcripts from conversations between student teachers in South Africa and the transcripts from previous research (Sjöberg, 2020; Sjöberg & Nyberg, 2020), to find excerpts where the student teachers reflect on teaching experiences as they relate to our understandings of ESD, PCK and the Anthropocene. Tentative themes were produced, each theme covering one aspect of the material only (Schreier, 2014). In the second phase of the analysis the authors revisited and reanalysed the data and finally came up with three stable themes: *Holistic understanding, Relevance of learning to everyday life* and *Modelling sustainable attitudes and behaviour.*

### *Student teachers' professional reflective conversations*

Although the data was drawn from distinct studies three common themes emerged in all the student teacher conversations. Despite the different contexts the student teachers agreed on three needs they considered important to becoming a STEM teacher. From the onset, the authors of this chapter from the two different settings were interested in the common themes arising in these conversations, as opposed to the differences between countries. These commonly articulated needs are: the need for holistic understandings of ESD and STEM teaching; the need for STEM teaching to be relevant to everyday life; and the need for STEM teachers to model sustainability attitudes and behaviours (Figure 6.1).

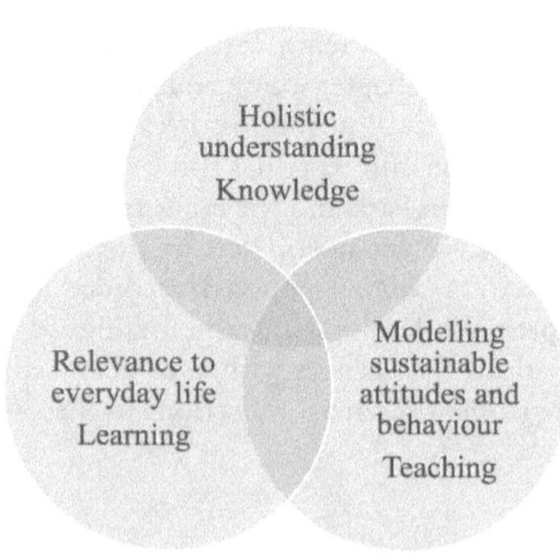


*Figure 6.1* Emergent themes in professional conversations

### *Holistic understanding – knowledge*

This theme reflects the type and character of the knowledge needed when teaching ESD in the STEM subjects. Student teachers expressed that it was not possible to teach ESD from one subject-specific viewpoint. One of the Swedish student teachers expressed this as a concern that not having an understanding of "what is included in all subject areas....it is difficult to know exactly what to focus on" (Sweden interview 1 – our translation) to teach effectively.

The conversations in both countries showed the student teachers think it is necessary to integrate different subject perspectives, but also importantly to plan for learning that negotiates, situates and integrates the school, the community and the world. The student teachers understood that working with sustainability and environmental problems demanded collaboration within the school and outside school to make students aware of the complexity of sustainable development. This was illustrated by one of the South African student teachers who explained:

> What I have learned is that, sometime to solve environmental problem, does not solely depend on an individual so it is a collaborative thing that you have to do, teachers, parents learners as well as the community members, the department as well as the government. So by working together I believe that the environmental problems can be solved and this sustainability programs to be a success, there should be a collaboration between the people I have mentioned...
>
> (SA Interview 3)

Particularly when talking about teaching ESD, as opposed to STEM, the student teachers expressed the need to leave the school in order to gain broader understandings of sustainability than can be taught in classroom. A Swedish student teacher explained that "the students have to understand quite complicated relationships ...in order to grasp the whole.... to understand the scale level you are at" (Sweden interview 3). A sentiment similar to that expressed in one of the South African conversations where it was explained that "sustainability education is not only in certain subjects like Geography and Life Sciences, it is a broad thing that involves many subjects and all members of society" (SA Interview 3).

This desire to open the classroom, to look outward for their students to understand the scale of sustainability, was repeated in the conversations, with special emphasis on collaborating with people in their local communities as is illustrated by a student teacher in the South African group interview who said, "once I get into the field and start teaching sustainable development, I should also integrate and involve other stakeholders and then embark on a common project as a community" (SA Interview 3).

### *Relevance of learning to everyday life – learning*

The second theme relates to what student teachers find essential for creating a classroom and school environment that offers opportunities for students to learn about ESD in the STEM subjects. During the conversations the student teachers recount that they feel it is important to use common, everyday examples when talking to their students about sustainability, and that this provided opportunities to use PCK. One of the South African student teachers highlighted this:

> another thing is that when teaching a lesson I will not start with the usual way of introducing the topic, instead I will pick a common relevant problem and present it to them for a solution, for example in rural areas many people use wood for fire to cook, meaning that if they continue to chop trees and burn wood, they are increasing the amount of carbon dioxide especially around their area. So I would if they find that a right thing to do and why or why not? How does it threaten their environment and contribute to global warming in general? From their response I will build on what they know
>
> (SA Interview 3)

While each conversation in South Africa mentioned this need to make learning about ESD more relevant to everyday life it was also viewed as important to keep students engaged and interested. The student teachers emphasised the importance in taking the students view as a starting point. One student teacher explained that when they started teaching full time they "would use more real life examples for learners to understand. Giving learners the platform to share their understanding to their classmates" (SA Interview 2). One of the Swedish conversations followed similar lines but added the following caveat:

> The question is how to handle that balance between simplification, but it must not be too simple either, so that some people lose interest, because they think it becomes too abstract.
>
> (Sweden interview 4 – our translation)

This question of balance comment also shows that student teachers felt a sense of responsibility to plan lessons that were learner centred and practical and that such lessons would translate into clear relevance for their students. A Swedish ST explained that they "have been thinking that [practical activities] should be on the central content of [STEM classes]... that this is what you should focus on teaching, and not on the knowledge requirements" (Sweden interview 2 – our translation). This aligns with a comment in one of the South African conversations:

> I would say this is a good way to start a lesson, and that is something I will have to sustain as a teacher. That is what I will take from here, teach a more learner centred lesson and bring in activeness because sustainable development topic requires a hands-on approach and experiential learning, I will further use Problem Based Learning throughout my lessons, like bring case studies for learners to engage in.
>
> (SA Interview 3)

However, it was also recognised that this kind of planning is not always easy:

> This, I feel is hard. To feel, that maybe not everyone is with you. You know that, but still you can hope they are. But, just try, I do not know, to keep up a sufficient pace to also engage those who are interested … and still not leave behind those who don't really follow. To not make them feel totally left behind and just feel: "I'll see what's happening on Facebook instead." This is difficult.
>
> (Sweden interview 8 – our translation)

*Modelling sustainable attitudes and behaviour – teaching*

In this third theme ideas about teaching come into focus. The student teachers described the role of the STEM teacher as being not only important for developing understanding and engagement but also as a model of attitudes and behaviours towards sustainability learning. The student teachers, particularly in the South African conversations, believed that it was important that "teachers we should lead by example, I should take the lead and pick what I see lying around as litter so that they can see that it's not for the learners only but also teachers should do it" (SA Interview 3). This particular conversation also clearly linked ideas about PCK to ESD learning. Building on the comment about teachers picking up litter in the school another student explained that:

> Us as teachers should go out there with learners and volunteer to pick litter for recycling purposes. We can also work with the learners to design poster and learning materials from these as well as post the message around school for everyone to adopt and sell the ideas on sustainable development.
>
> (SA Interview 3)

## Discussion

Three themes emerged from the content analysis of the professional reflective conversations between student teachers and teacher educators in the teacher education programmes in South Africa and Sweden about teaching in the Anthropocene. The reflections include the need for student teachers to: have holistic understandings of ESD and STEM teaching; understand and incorporate how STEM is relevant to everyday life; and model sustainability attitudes and behaviours to the school students they teach. Because in each case the conversations were conducted with a teacher educator present, and with a focus on teaching practice experiences, it is not surprising that the themes align with knowledge and understanding, learning and teaching. What is interesting is that in these conversations, the student teachers were able to reflect on what they had done and how they might do things differently in relation to PCK and bringing ESD to the fore for the students.

> *Research Question 1– In what way do professional conversations between student teachers and teacher educators allow for, and promote reflection on teaching experiences in the Anthropocene?*

These two studies were different, in terms of focus and purpose which led to many different types of conversations. However, each conversation allowed similar possibilities for reflection and in certain ways provided time and space that made the student teachers reflect on their teaching experiences. In the conversations, student teachers reflected on both specific situations and more generally about teaching practice and the role as teacher in this situation. The conversations themselves were viewed as valuable learning experiences for the student teachers and helped them develop their teaching practice. The following excerpt shows that student teachers were able to make conclusions about the impact of being able to talk about this:

> Yes, this conversation made me realise I was doing everything wrong; listening to everyone's responses, they have use different strategies for learners whereas I was teaching the traditional way – the way I was taught without realising that the times have changed. And, I really did not understand when my teachers were teaching me yet I was doing the same methods and hoping learners to understand.
>
> (South African Interview 2)

Here we see that when they are given opportunities to have professional and reflective conversations with teacher educators on teaching practice the student teachers felt the gap between theory and practice is somewhat diminished.

> *Research Question 2 – What new insights are gained between teacher educators/researchers from the north and south regarding new ways of developing and facilitating PCK and ESD in science teacher education?*

One of the insights that emerged from the conversations was that student teachers felt it important for their students and other teachers to be aware of the complexity of sustainability issues. While not explicitly mentioned the conversations recognised the responsibility of teachers to work towards the Sustainable Development Goal – 4 which calls for sustainability to be integrated into teaching and learning practices:

> By 2030, ensure that all learners acquire the knowledge and skills needed to promote sustainable development, including, among others, through education for sustainable development and sustainable lifestyles, human rights, gender equality, promotion of a culture of peace and non-violence, global citizenship and appreciation of cultural diversity and of culture's contribution to sustainable development
>
> (UNESCO, 2017; Goal 4.7)

The complex notion of ESD provided the student teachers different entry points into the professional reflective conversations. As a South African participant explained, "I have also learned [in this conversation] that this sustainability education it is a broad thing that involves many subjects and all members of society" (SA Interview 3).

Building on this the student teachers felt that being part of focused conversations with teacher educators and other student teachers enabled them to be exposed to other views, which in turn encouraged them to think of other ways of teaching and contributed to the development of PCK and explicitly bridge the theory practice gap. The student teachers explained that all teachers, not just new teachers like them, require greater PCK in the Anthropocene which includes the toolkit of diverse strategies. Having the opportunity to discuss diverse strategies was valued. One of the Swedish student teachers explained that this was important because the learners in the classroom were also diverse and conversations could help with "ideas of how you teach all seventeen different students… But when it is different content, [learning other ways] fits well when teaching students of different abilities" (Sweden meeting 2). Sharing diverse strategies and questioning the teaching strategies used by others in conversation means that they are more aware of other perspectives and they are more prepared.

However, in both cases, the conversations changed quite quickly from sharing teaching or instructional strategies to focus more on student learning. This was evident in the widespread understanding that teachers have a responsibility to ensure ESD teaching and learning was relevant to the everyday life of their students. The South African conversations focused on how PCK could inform the planning of lessons that make sustainability relevant and ensure students learned that ESD is about them living on this planet and is important for the future. The Swedish conversations also recognised the complexity of sustainability issues, but tended to focus on how their own PCK could better tell them about student understanding and as a framing for what teacher education should aim for.

## Conclusion

The collaboration and indeed deep conversations about teacher education for this chapter allowed us to look more deeply at how teacher education programmes in the two countries provided opportunities for student teachers in STEM to discuss their developing PCK. The student teachers recognised in dialogue with teacher educators that ESD was complex and they became increasingly aware of the need for teachers to integrate content from different areas. Taking that one step further, there was also recognition that to bring these complex ideas to the classroom they needed to transform the content knowledge they held into PCK.

We found that outside of the school practice periods there are few official opportunities in the teacher education programmes in both countries to have structured reflective conversations as part of the degree. This indicates that the value of professional reflective conversations is underestimated in the teacher education programmes outlined in this chapter. However, when student teachers are given the time, space and knowledge to engage in dialogue with their peers and teacher educators the PCK needed to teach STEM subjects is enhanced. To be able to successfully navigate teaching STEM in the Anthropocene, we believe teacher education programmes should provide more opportunities for student teachers to have deep reflective conversations focusing on the development of PCK.

## Note

1. In Sweden and other Scandinavian countries together with Germany, the term *didaktik* is often used to encompass the science of the teaching profession and the professional content knowledge (PCK) that teachers possess (Kindenberg & Wickman, 2018; Osbeck, Ingerman, & Claesson, 2018).

## References


Appleton, K. (2013). Science pedagogical content knowledge and elementary school teachers. *Elementary science teacher education: International perspectives on contemporary issues and practice* (3rd ed.). New York, NY: Routledge.

Brandt, J. O., Bürgener, L., Barth, M., & Redman, A. (2019). Becoming a competent teacher in education for sustainable development. *International Journal of Sustainability in Higher Education, 20*(4), 630–653.

Brennan, M. (2017). Struggles for teacher education in the age of the Anthropocene. *Journal of Education, 69*, 43–66.

Clarke, A., 1995. Professional development in practicum settings: Reflective practice under scrutiny. *Teaching and Teacher Education*, 11, 243–261. doi: 10.1016/0742-051X(94)00028-5

Crutzen, P. J. (2002). Geology of mankind. *Nature* 415, 23. doi: 10.1038/415023a; pmid: 11780095

Eckerman Pitton, D. (2006). *Mentoring novice teachers: Fostering a dialogue process* (2nd ed.). Thousand Oaks, CA: Corwin Press.

Evans, N. (2019). Teacher education and education for sustainability. In J. Ferreira, N. Evans, J. M. Davis, & R. Stevenson (Eds.), *Learning to embed sustainability in teacher education.* Singapore: ISBN: 978-981-13-9536-9 (eBook). New York, NY: Springer.

Graneheim, U. H., & Lundman, B. (2004). Qualitative content analysis in nursing research: Concepts, procedures and measures to achieve trustworthiness. *Nurse Education Today, 24*(2), 105–112.

Gravett, S., de Beer, J., Odendaal-Kroon, R., & Merseth, K. K. (2017). The affordances of case-based teaching for the professional learning of student-teachers. *Journal of Curriculum Studies, 49*(3), 369–390. doi: 10.1080/00220272.2016.1149224

Grossman, P., Schoenfeld, A., & Lee, C. (2007). Teaching subject matter. In L. Darling-Hammond, & J. Bransford (Eds.), *Preparing teachers for a changing world: What teachers should learn and be able to do.* San Francisco, CA: Jossey-Bass.

Hipkiss, A. M., Windsor, S., & Sanders, D. (2020). The girl with the garden gloves: Researching the affordances of sensual materialities in the school garden. *Ethnography and Education, 15*(3), 350–362. doi: 10.1080/17457823.2019.1698309

Hsieh, H. F., & Shannon, S. E. (2005). Three approaches to qualitative content analysis. *Qualitative Health Research, 15*(9), 1277–1288.

Hume, A., Cooper, R., & Borowski, A. (Eds.). (2019). *Repositioning pedagogical content knowledge in teachers' knowledge for teaching science* (pp. 77–92). Singapore: Springer.

Kessels, P., & Korthagen, F. (1996). The relationship between theory and practice: Back to the classics. *Educational Researcher, 25*, 17–22.

Kindenberg, B., & Wickman, P. O. (2018). Dags för didaktiken att bli egen vetenskap. *Pedagogiska Magasinet*, 18(2), 14–17.

Kriewaldt, J., Nash, M., Windsor, S., Thornton, J., & Reid, C. (2018). Fostering professional learning through evidence-informed mentoring dialogues in school settings. In J. Kriewaldt, A. Ambrosetti, D. Rorrison, & R. Capeness (Eds.), *Educating future teachers: Innovative perspectives in professional experience*. Singapore: Springer.

Loughran, J. J. (2014). Developing understandings of practice: Science teacher learning. In N. G. Lederman, & S. K. Abell (Eds.), *Handbook of research on science education* (Vol. II, pp. 825–843). New York, NY: Routledge.

Mun Ling, L., & Marton, F. (2012). Towards a science of the art of teaching. *International Journal for Lesson and Learning Studies, 1*(1), 7–22.

Osbeck, C., Ingerman, Å, & Claesson, S. (2018). *Didactic classroom studies: A potential research direction*. Gothenburg: Kriterium.

Park, S., & Oliver, J. S. (2008). Revisiting the conceptualisation of pedagogical content knowledge (PCK): PCK as a conceptual tool to understand teachers as professionals. *Research in Science Education, 38*(3), 261–284.

Saric, M., & Steh, B. (2017). Critical reflection in the professional development of teachers: Challenges and possibilities. *CEPS Journal, 7*(3), 67–85.

Schreier, M. (2014). Qualitative content analysis. In U. Flick (Ed.), *The SAGE handbook of qualitative data analysis* (pp. 170–183). London: SAGE Publications Ltd. doi: 10.4135/9781446282243

Shulman, L. (1987). Knowledge and teaching: Foundations of the new reform. *Harvard Educational Review, 57*(1), 1–22.

Shumba, O., & Kampamba, R. (2013). Mainstreaming ESD into science teacher education courses: A case for ESD pedagogical content knowledge and learning as connection. *Southern African Journal of Environmental Education, 29*, 151–166.

Sjöberg, M. (2020). *Samtal om undervisning i naturvetenskap - Ämnesdidaktisk kollegial utveckling i lärarutbildning och lärarprofession* (translated title: *Conversations about science teaching experiences – Collegial learning and development of professional knowledge for science teaching in teacher education and the teaching profession*. Doctoral Thesis; Gothenburg Studies in Educational Sciences, Gothenburg University. http://hdl.handle.net/2077/64566

Sjöberg, M., & Nyberg, E. (2020). Professional knowledge for teaching in student teachers' conversations about field experiences. *Journal of Science Teacher Education, 31*(2), 226–244. doi: 10.1080/1046560X.2019.1688533

Sööt, A., & Viskus, E. (2015). Reflection on teaching: A way to learn from practice. *Procedia – Social and Behavioral Sciences, 191*, 1941–1946.

Steffen, W., Crutzen, J., & McNeill, J. R. (2007). The Anthropocene: Are humans now overwhelming the great forces of nature? *Ambio, 36*, 614–621.

Timperley, H. (2001). Mentoring conversations designed to promote student teacher learning. *Asia-Pacific Journal of Teacher Education, 29*(2), 111–123.

Timperley, H. S., & Alton-Lee, A. (2008). Reframing teacher professional learning: An alternative policy approach to strengthening valued outcomes for diverse learners. *Review of Research in Education, 32*(1), 328–369.

UNESCO. (2015). *Incheon declaration education 2030: Towards inclusive and equitable quality education and lifelong learning for all.* Retrieved from http://unescodoc.unesco,org/images/0023/002331/233137E.pdf

UNESCO. (2017). *Education for sustainable development goals: Learning objectives.* Paris, France: UNESCO.

UNESCO. (2019). Press release *UNESCO 40th General Conference adopts a new global framework for Education for Sustainable Development for 2020–2030.* https://en.unesco.org/news/unesco-40th-general-conference-adopts-new-global-framework-education-sustainable-development

United Nations (UN). (2015). *Transforming our world: The 2030 agenda for sustainable development.* Paris, France: United Nations.

van Kruiningen, J. F. (2013). Educational design as conversation: A conversation analytical perspective on teacher dialogue. *Teaching and Teacher Education*, *29*, 110–121.

Windsor, S., Kriewaldt, J., Nash, M., Lilja, A., & Thornton, J. (2020). Developing teachers: Adopting observation tools that suspend judgement to stimulate evidence-informed dialogue during the teaching practicum to enrich teacher professional development. *Professional Development in Education*, 1–15. doi: 10.1080/19415257.2020.1712452

# 7 Exploring a South African and a Swedish teacher-education programme of Biology teachers for ESD

## A comparative study

*Portia Kavai, Alexina Thorén Williams and Maria Tsakeni*

## Introduction

In the backdrop of the Anthropocene, which is the human-activity dominated epoch (Lewis & Maslin, 2015), sustainable development embodies efforts to control the impact that human activities have on the well-being of the planet that we inhabit. Several action plans for sustainable development have been put into place. Amidst concerted efforts towards the realisation of the Agenda 2030 through the attainment of the sustainable development goals (SDGs), this study argues that one of the pathways to sustainability is to prepare student teachers to implement education for sustainable development (ESD) in their future classrooms. Sequel to the realisation that development should meet the present demands for all people and future generations, a number of actions have been taken at a global level. These include the pronouncement of the decade dedicated to ESD (2005–2014), the millennium development goals (MDGs) and the SDGs as global action plans (Muralikrishna & Manickam, 2017; Woodridge & Zimmermann, 2015). The decade for ESD lasted between 2005 and 2014, and the millennium development goals agenda and implementation were between the years 2000 and 2015 (Cuenca-García, Sánchez, & Navarro-Pabsdorf, 2019). With the lapsing of the timeframes for the two sets of development goals came the pronouncement of the 17 SDGs through the United Nations Agenda 2030 (United Nations, 2018). Therefore, SDGs are a build-up of their predecessors, namely, sustainable development, ESD and the millennium goals. One model of sustainable development describes development to be comprehensive by ensuring social, economic and environmental well-being (Muralikrishna & Manickam, 2017). However, specific action plans are required to achieve both sustainable development and ESD. For instance, the millennium development goals targeted specific issues identified as challenges encountered by human populations. The eight millennium goals aimed to eradicate extreme poverty and hunger; achieve universal primary education; promote gender equality and empower women; reduce child mortality; improve maternal health; combat

DOI: 10.4324/9781003190158-10

HIV/AIDS, malaria and other diseases; ensure environmental sustainability; and develop a global partnership for development. The indicators for sustainable development expanded into the 17 SDGs espoused by the United Nations Agenda 2030 (United Nations, 2015). The SDGs build on the millennium goals (Woodridge & Zimmermann, 2015). There are claims that the millennium goals have reduced world poverty (Muralikrishna & Manickam, 2017; Woodridge & Zimmermann, 2015). However, there is still a lot that needs to be done judging by the target areas for improvement defined by the SDGs. Implementation of the SDGs faces challenges such as financing, collection of data for monitoring the absence of a holistic as well as an integrated approach towards set objectives (Woodridge & Zimmermann, 2015). Therefore, the development of effective implementation methods to ensure the success of the SDGs is one of the challenges facing the sustainability agenda. ESD is one of the approaches that can be used to achieve the SDGs. ESD is concerned with finding models to integrate sustainable development values, practices and principles in formal and informal education. Our interpretation is that ESD's ultimate goal is to prepare citizens who have the appropriate skills and values to act responsibly, guided by the principles and goals of sustainable development (Schild, 2016).

An approach towards introducing sustainable development in education has been to integrate the concept across school curricula. Each school subject has a unique part to play in the teaching of ESD. Accordingly, science subjects such as Biology are strategically positioned to promote some of the SDGs through education, based on the overlap and integration of content. Therefore, it might be easy to identify some SDGs that have broadly overlapping concepts in Biology. For instance, SDG numbers 2, 3, 4, 6, 12, 13, 14 and 15 (see Appendix) can be considered as some of the topics existing in Biology as a discipline (United Nations, 2018). These topics can also be contained as interdisciplinary concepts in other science disciplines such as Geography, Physics and Chemistry. The topics include food security, improved nutrition, sustainable agriculture, healthy lifestyles, sustainable water management and sanitation. Climate change, sustainable consumption and production of materials, sustainable use of oceans, seas and marine resources, and sustainable ecosystem management are some of the topics.

Holbrook (2009) explains that Science education plays a part in the fulfilment of SDGs because the subject aims to equip learners with the ability to use the acquired knowledge and skills in everyday life by making informed decisions to improve the quality of life. The SDGs aim to improve the quality of different aspects of life, the state of the environment and how to achieve the desired sustainable "doings" and "beings". That is the capability of undertaking the actions and activities needed (Robeyns, 2005) for the well-being of our society, the entire global community and the biosphere. Incorporating the SDGs in Science education also improves the relevance of the discipline to learners' lives.

The pronouncement of the SDGs is a fairly recent development, and one wonders how the teaching of Biology is currently contributing to the achievement of the goals. The puzzle further extends to how Biology teachers are prepared to teach for sustainable development. It has been established that Biology teachers struggle with ESD integration due to their insufficient knowledge and skills. Karaarslan and Teksöz (2016) observe that simplistic approaches of integrating ESD into the curricula of Science and Technology teacher education programmes are not enough to ensure that the teachers possess adequate knowledge, competencies and skills. It is important that teacher education programmes practically ensure that competencies of student teachers in science subjects are related to the essential characteristics of ESD, holistic approach, envisioning change and achieving transformation (UNECE, 2011). Therefore, in this study, we explored Biology teacher education with regard to how student teachers are prepared for ESD based on the global sustainable development imperatives such as the SDGs in a comparative case study of a South African and a Swedish university.

## The role of Biology in sustainable development

One of the SDGs (SDG no. 4, see Appendix) emphasises the importance of inclusive, equitable and quality education with improved lifelong learning opportunities (United Nations, 2018). It would be interesting to see how student teachers are prepared to ensure that Biology education encapsulates the values of inclusivity and equitable access in addition to providing learners with lifelong learning opportunities. As pointed out in the previous section, some of the SDGs exist as topics in Biology as a Science subject. However, the teaching of the topics may not translate to the teaching of ESD and the attainment of the objectives of the SDGs as espoused. Grimm (2008) observes that some practices in education have contributed to the current sustainability challenges. Grimm (2008)'s observation implies that Biology should be taught for sustainable development, and the existence of common topics between the subject and the SDGs serves as an enabling springboard for implementation.

Some studies on the integration of sustainable development into Science teaching use environmental education as an entry point. The assumption is that environmental education focuses on the inculcation of behaviours, values and practices for the sustainability of life on planet earth (Schild, 2016; Stanisic & Maksic, 2014). The environment is one of the major pillars for sustainability in addition to the social and economic pillars (Burmeister, Rauch, & Eilks, 2012). Abdullah, Halim and Shahali (2011) observe that environmental education is integrated more in Biology than the other Science subjects of Chemistry and Physics in the Malaysian context. Despite the importance of the environment to the whole concept of sustainability, holistic approaches to the implementation of ESD through school subjects should be explored and implemented.

## The five-pillar framework of education for sustainable development

Sustainable development is pronounced at a global level by policymakers from different countries and guided by world bodies such as the United Nations.

Sustainable development is a way of rethinking development to meet the needs of present societies without compromising the needs of future generations (Borg, Gericke, Höglund, & Bergman, 2014).

There are implications for sustainable development in every sphere of life, including education. The Declaration of Thessaloniki (UNESCO, 1997) gives some recommendations for sustainable development in education, including sustainable development expectations for schools and teacher education. First, it recommends that schools be supported as they develop curricula that encompass sustainable development. Second, the declaration (UNESCO, 1997, p. 3) says the following about teacher education: "Special emphasis should be given to strengthening and eventual reorientation of teacher training programmes and identification and sharing of innovative practices. Support should be given to research in interdisciplinary teaching methodologies and assessing the impact of relevant educational programmes". Besides, there is a recommendation to reorientate education as a whole (formal and informal) towards sustainability in all countries. Out of the endeavours to incorporate sustainable development objectives as part of pedagogy, the term education for sustainable development was coined. Accordingly, ESD is a teaching approach for learner development in promoting sustainable development (Borg et al., 2014).

To understand ESD, it is important to work with a sustainable development model. The most popular sustainable development model seeks to foster sustainable development in the areas of environment (ecology), society and economics (Burmeister et al., 2012). According to this model, sustainability is maintained by three pillars: *economic*, *social* and *environmental*. Some authors have made attempts to operationalise ESD implementation in Science education based on this sustainable development model and other factors. Eilks (2015) mentions four models of integrating ESD in Science classrooms. The models are (i) adopting principles from sustainable practices in Science and Technology for hands-on Science education laboratory work; (ii) adding sustainable science as content in the Science and Technology curriculum; (iii) using controversial sustainability questions for the socio-scientific issues driving Science education; and (iv) Science education as a part of ESD-driven school development. This study used an ESD conceptual framework that consists of the three mentioned pillars of sustainable development and two additional pillars, which have been coined as the *citizen skills* pillar and the *pedagogies for ESD* pillar (Tsakeni, 2018). The conceptual framework is illustrated in Figure 7.1.

The model consists of the teacher knowledge components that will enable ESD implementation in the science classrooms. The ESD model indicates the knowledge components that student teachers have to develop in order to be able to implement ESD in their future classrooms (Tsakeni, 2018). Tsakeni (2018) justifies the addition of the *citizen skills* pillar because ESD is about developing values and ethics that inform young people to behave in responsible ways through participatory learning in formal and informal settings. The education component of sustainable development speaks to how teachers facilitate the process of teaching and learning to achieve the objectives of ESD. Therefore, the

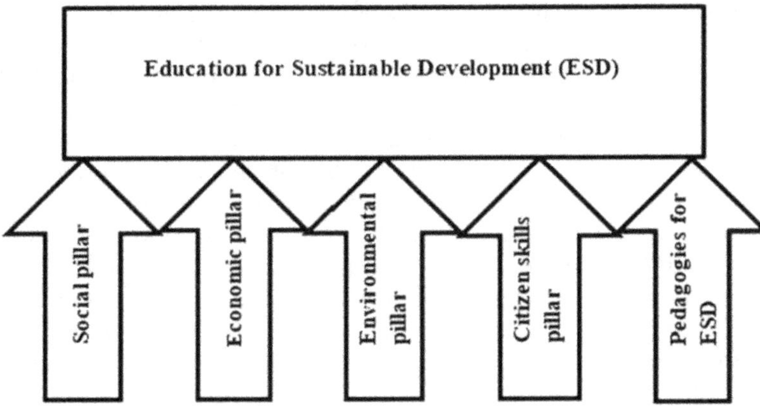


*Figure 7.1* Education for sustainable development conceptual framework

Adapted from Maria Tsakeni (2018, p. 127) Opportunities for teaching sustainable development through the chemistry component of CAPS physical sciences, African Journal of Research in Mathematics, Science and Technology Education, p. 127, copyright © SAARMSTE, reprinted by permission of Taylor & Francis Ltd, http://www.tandfonline.com on behalf of SAARMSTE.

*pedagogies for ESD* pillar also stand as a critical component in this study as it looks at how student teachers are being prepared for practice in Biology classrooms. The ESD model, developed by Tsakeni (2018,) is used as an analytical framework to elicit from the data collected on how the two teacher preparation programmes prepare the Biology teachers in the identified knowledge components.

## Research questions

The study explored how prospective Biology teachers are prepared for ESD by comparing practices between a South African and a Swedish university. The main research question is: How do teacher education programmes prepare prospective Biology teachers for ESD in a South African and a Swedish university?

The sub-research questions are:

i How do policies guide Biology student teacher preparation for ESD at a South African and a Swedish university?
ii According to Biology teacher educators, how do they prepare student teachers for ESD at a South African and a Swedish university?

## Context of study and data collection

### *Biology (Life Sciences) teacher education programme in South Africa*

In the South Africa education system, Biology is referred to as Life Sciences. From the third year of the four-year teacher education programme for secondary school, students within science choose one area of specialisation: Life Sciences (Biology) or Physical Sciences (Physics and Chemistry). The secondary school phase, which these student teachers are trained to teach, is named the Further Education and Training (FET) (Grades 10–12). The ecological issues form an integral part of both Life Sciences (Biology) teacher education and high school Life Sciences in South Africa. It emphasises ecosystems, biodiversity, life cycles of plants and animals and life-supporting processes such as photosynthesis and respiration.

International agreements on sustainable development guide countries as they develop qualification frameworks for teacher education. Among the policies that guide teacher education in South Africa is the Minimum Requirements for Teacher Qualifications (MRTEQ), developed by the Department of Higher Education and Training (Department of Higher Education and Training (DHET), 2011).

### *Biology (Life Sciences) teacher education programme in Sweden*

In Sweden, the education for Biology teachers in secondary school is a five-year Special Subjects Teacher Programme (300 credits), which corresponds to the South African FET phase. In order to teach Biology in a secondary school in Sweden, student teachers can be either specialised to teach in the lower secondary school (Grades 7–9, students age 13–15) or to teach in upper secondary school (students age 16–19). In either case, and if they take Biology as a Major (90 credits), they need to complete at least three compulsory basic courses in Biology (45 credits), including subject-specific didactics (7,5 credits). In addition to these courses, they need to take two more courses (each 15 credits), of which one is selectable (e.g., nature conservation in aquatic environments, molecular microbiology, genetics, human physiology, evolutionary biology) and the other compulsory. The focus in the compulsory course is on subject-specific didactics (7,5 credits) and project in Biology (7,5 credits). The three basic courses are (i) structure of organisms, genetic variation, evolution, reproduction and diversity, (ii) photosynthesis and combustion, (iii) anatomy and function of vertebrates. The sub-course in subject-specific didactics includes Science education in an international and national perspective and ESD in Biology education. The programme ends with a degree project at the advanced level. The teacher educators who teach these courses have a background as scientists within a particular field of Biology or as schoolteachers with a doctoral degree in Science education.

In Sweden, the policies that guide the teacher education programme are the *Special Subjects Teacher Programme Syllabus*, issued by the Gothenburg University Coordination Board for Teacher Education, and the *Subject Guide for Biology* within the Special Subjects Teacher Programme. The Department of Biological and Environmental Sciences issues the latter document. Taken together, these minimum requirements guide the design and development of qualifications for teachers and other professionals working in the education sector in Sweden.

### *Data collection*

The study is a comparative qualitative case study. A case study can be defined by a phenomenon of interest that calls for research (Creswell, 2007). The phenomenon under study is how teacher education programmes prepare prospective Biology teachers to teach sustainable development by comparing practices at a South African and a Swedish university. Data were collected by means of document analysis of the policies that guide Biology teacher education at the two universities. In South Africa, the MRTEQ, the university handbook and the secondary school study guides for Biology student teacher preparation were analysed for content. Similarly, at the Swedish university, the syllabus for the Special Subjects Teacher Programme directed towards upper-secondary school and the *Subject Guide for Biology* within the Special Subjects Teacher Programme were analysed. Semi-structured interviews were also conducted with two Biology teacher educators at each of the universities. The criteria for selecting the teacher educators in South Africa and Sweden were based on whether they were involved in preparing Biology teachers, their commitment to ESD and their availability. In the Swedish case, the selection of teacher educators was particularly dependent on the teacher educators' availability and willingness to participate in the interviews. In the South African case, it was central to select teachers who could provide rich data and had substantial ESD backgrounds.

The interview questions were grouped into three themes. These were: (i) teacher educators' experiences with ESD, (ii) teacher educators' pedagogies for ESD and (iii) outcomes of the ESD student teacher preparation. Although the interview questions were structured according to these three themes, they were open-ended. Each teacher educator was interviewed for about an hour.

### *Data analysis*

The data collected under three themes generated teacher educator narratives. The narratives and data from document analysis were analysed for content with the research questions in mind in addition to the five pillars of ESD contained in the conceptual framework in Figure 7.1. The thematic analysis was combined with constant comparison techniques analysis to bring the two data sets together and enable the comparison of the two teacher education practices for preparing prospective Biology teachers for ESD.

## Findings of the study

### ***ESD in Biology teacher education – South African case***

The participating teacher educators at the South African university were assigned the pseudonyms South African university lecturer 1 (SAL1) and South African university lecturer 2 (SAL2). SAL1 was a 49-year-old male teacher educator teaching Life Sciences (Biology) and methodology courses to prospective

Biology teachers. SAL2 was a 40-year-old female teacher educator teaching Life Sciences and methodology courses to the same category of student teachers.

### *South African policy documents guiding teacher education for ESD*

The environmental pillar of the ESD conceptual framework in Figure 7.1 is strongly supported by the topics under the ecological dimension of the Biology curriculum. In the context of teaching and learning of sustainable development, the ecological dimension refers to the ecosystems that include all living things, resources and life-supporting systems. Its goal is to promote the conservation of these ecosystems in the Anthropocene era. This dimension is quite visible in the policy documents, including the UNECE (2011), the Department of Education, National Curriculum Statement (2011), the university handbook and the course modules. One aspect that is common in all these documents is that there is no clear or explicit objective dealing with sustainability perspectives. Upon reviewing the objectives for each of the Life Sciences' third- and fourth-year courses, it was evident that sustainable development integration was not clearly outlined in any of the course objectives. It was more implicit than explicit to a point where one may have to deduce or suppose the instances where sustainable development can be discussed.

The South African participating teacher educators were asked about other policies that inform the integration of sustainable development. Interestingly, they seemed to be of the opinion that there was no policy guiding teacher educators' integration of sustainable development except, according to SAL1, "own agency and if the teacher educator has values, knowledge, attitudes, skills to find ways of implementing ESD in their own classrooms". SAL1 went on to say: "Considering the South African Constitution on the right to a clean and a healthy environment can be a starting point that one has to think of and then be deliberate on 'How can my subject contribute to a healthy environment? How can I and my discipline contribute?' SAL2 was not sure if the policies existed or not and referred the interviewer to another colleague known to be strong on policies. The colleague referred to was SAL1, who had already alluded to the non-existence of such policies. The only document in place in the institution was the ESD module, which was developed for the master's degree programme. Although the ESD module had already been approved, it was yet to be implemented. It was hoped that it could be taught in the following year (2020) if there would be personnel confident enough to teach it. Therefore, according to the policies, content that could support the environmental pillar of ESD was present. What was not clear was the articulation of pedagogies for ESD. Besides, SAL1's reference to "own agency" seemed to point to the notion that lecturers also needed to have developed citizen skills and pedagogical skills for ESD to function well in the area.

### *South African teacher educators' views*

Both South African participants (SAL1 and SAL2) acknowledged that they were familiar with the millennium goals and SDGs and had developed knowledge and beliefs acquired through previous work experiences. SAL1 revealed that he was familiar with the SDGs since he had had previous work experiences as a trainer in the millennium goals, with a specific focus on environmental education and climate change. He said:

> Very familiar with SDGs: I have worked in the field of environmental education since 2002 related to millennium development goals. In 2005, a decade of education for sustainable development still in environmental education was launched. Environmental education … related to ESD.

The assumption is that SAL1's experiences would influence his knowledge and beliefs as he worked in the preparation of Biology teachers. It seemed the focus of his familiarity with ESD was through environmental education and some developmental and socioeconomic issues derived from the millennium development goals. SAL1 further said:

> Environmental education [in Africa] is not viewed as the European one because the European one is more of nature education. But here in this continent, it is more socioeconomic, biophysical and social factors … environment of human being. Education is called to take up that role to deal with developmental concerns; how to deal with change in the environment; how education can respond to those changes. But the human component was not foregrounded. It is more of the biophysical environment, not the human.

SAL2's experience with the concept of sustainable development was through two modules that she had participated in developing. One of the modules was on sustainability, and the other was on ESD. On being asked how familiar she was with the SDGs, SAL2 said:

> Pretty familiar. Millennium sustainability and other issues. I participated in developing a master's programme module on sustainability, not necessarily for teacher educators, and another module that is specific for sustainability in education.

SAL2 seemed to understand sustainable development through the three-pillar (social, environmental and economic) model. However, she emphasised the social and environmental pillars. On being asked what she understood by the term sustainable development, she said:

> Social issues and environmental issues and the problems we are trying to find solutions to. Sustainability issues vary according to the context. The outcome is to explain the meaning of sustainability; the interplay between the components of ESD, social learning, the environmental and social pillars. Looking at all the pillars is feasible depending on who the candidates are.

The findings in this section seem to suggest that the two participating South African teacher educators possess varied experiences, knowledge and beliefs regarding the concept of ESD that they use as a filter as they prepare future Biology teachers.

Despite pointing out that the policies were not very specific on how to teach ESD to student teachers, SAL1 and SAL2 described instructional methods that they used in the classroom. SAL2 thought that class discussions could be exploited as an integrated platform by deliberately stirring conversations to include ESD issues. She said:

> I stir the conversation on how we unpack sustainability in whatever we are teaching in almost every topic, especially with my PGCE students. It's kind of something that ESD goals must be linked in every topic; therefore, I encourage sustainability conversation in every topic.

However, SAL2 felt that it was difficult to incorporate ESD into some Biology-content topics taught for conceptual understanding. She also indicated that there were some topics in which it would be easy for her to include ESD. She said:

> Some topics like genetics are difficult to incorporate ESD because they are taught conceptually, but when I link genetics and evolution, then I can refer to sustainability. With topics like biodiversity and ecology, it's much easier to incorporate ESD.

On the same note, SAL1 seemed to support the notion that most Biology-content courses were taught to achieve other objectives, such as conceptual understanding when he observed that ESD seemed to be integrated more in the methods courses. He said: "Methods sections then take up the responsibility to teach ESD, but it would be more effective if it was linked to content lectures."

As SAL1 further explained, the incorporation of ESD into the teaching and learning process heavily depended on the capabilities and abilities of the individual lecturer. He said:

> Course outlines just have a title like biodiversity and the environment or population ecology and environment … As a lecturer, one becomes aware of space and aware the importance of the environment in which population ecology is taking place, but it is not explicit.

SAL1 thought that ESD should be taught through strategies in which the student teachers actively participate in constructing the knowledge if they are to be effective practitioners in their future classrooms. He said:

> For my students, most methods may work, but especially participatory methods, so that they start using the knowledge in participation. Acquisition through participation. It is important to have a balance so that when they

acquire the agency, they engage, have capacity to function and participate and link it to the topics they will teach in their schools.

In an attempt to illustrate how he thought ESD could be taught to student teachers in authentic ways, SAL1 mentioned some real-life scenarios that can be brought into the teaching and learning situation. He also emphasised the importance of including formative and summative assessments. He said:

> It is difficult to instil ESD in large groups; so far, I think one could think of a task I would give. Would love to do an audit like taking a bin from outside and take an audit of that bin and conscientize them [student teachers] and relate with the knowledge and practices in their lives ... formative and summative assessment, out-of-class activities, projects.

The two teacher educators believed that the prospective Biology teachers developed skills and a value system as they are prepared for ESD classroom practice. SAL2 seemed to suggest that the student teachers develop some positive values and pedagogical skills after going through teacher preparation, as could be gleaned from the following interview excerpt.

*RESEARCHER:* Can they [student teachers] acquire the ESD values and then pass it on to their students?

*SAL2:* Yes, because we have highlighted it in class and they are more aware of these shifts and the methodologies. We keep on telling them you can't change the education system but make a change in your own classroom. This group of students can see the social and environmental issues, they can see that they have the capacity to make a change.

Similarly, SAL1 observed that some of his students might have developed some positive attitudes towards environmental sustainability, which could lead to some action-taking citizen skills and pedagogies for ESD in their future classrooms. He said:

> When they did the task, they did not appreciate it at that time, but as they are walking in the streets, they now take notice of what is happening in the environment around them, for example, sewage spills and implications on the environment, which may cause diseases. Some even consider options like reporting to the municipality, write a letter to the municipality authorities. Conscientizing [the student teachers] helped them apply ESD to real-life scenarios that they can pass on to their learners.

The two South African teacher educators brought to the fore previously acquired experiences and knowledge of ESD that have influenced how they prepare future Biology teachers. The environmental, social, citizen skills and pedagogies components of ESD were more visible than the economic component.

This observation was made by SAL1 when asked if the economic dimension of ESD was treated: "I don't think it is well reflected; some are reflected, but if I think about it, we should do more for it to be reflected."

### *ESD in Biology teacher education – Swedish case*

The participating teacher educators at the Swedish university were assigned the pseudonyms Swedish university lecturer 1 (SWL1) and Swedish university lecturer 2 (SWL2), respectively. SWL1 was a 61-year-old male lecturer teaching Physiology to future Biology teachers. SWL2 was a 67-year-old female lecturer teaching Science Didactics to future Life Sciences teachers. The findings are organised in two major sections according to the two research questions of the study and discussed below.

#### *Swedish policy documents guiding teacher education for ESD*

Similar to the SA policy documents, Swedish policy documents emphasise the ecological dimension of learning for sustainable development. However, this was quite evident for the first two basic Biology courses for prospective secondary Biology teachers. In the third course, the objectives are related to the social dimension of sustainable development. There is also a difference between the first and second Biology courses. In the first course, the ecological dimension of sustainable development is implicitly articulated through the objectives "... to have knowledge about the biome of the world; to understand the impact of humans on these; to have the ability to evaluate biodiversity from a scientific perspective". In the second Biology course, the ecological dimension is explicit through the objectives "... to demonstrate knowledge of the research field: Education for sustainable development; to problematize issues related to natural resources, climate change, and other aspects of Sustainable Development". In the third course, which is about Biotechnology and the shape and function of vertebrate animals, the social dimension of sustainable development is articulated through the objectives of social and ethical aspects of biotechnology. The ecological and social dimensions of sustainable development are explicitly emphasised in the elective courses on the conservation of nature in aquatic environments, microbiology and evolutionary biology. In the case of Sweden, two dimensions of ESD, environmental and social, are clearly visible.

Sustainable development is included in the Biology courses as an underlying issue that should be integrated into broader content topics in Biology. The inclusion of ESD is both explicit and implicit. In cases where the courses articulate broad ESD objectives, instructional strategies are not suggested. Accordingly, it can be assumed that issues such as animals at risk of extinction, pollution and eutrophication could be discussed under broader topics such as ecology, ecosystem and biodiversity. Some other topics under which sustainable development issues could be mentioned are life cycles of plants and animals, life-supporting processes such as photosynthesis and respiration and conservation issues.

The most salient ESD pillar in the course content is the environmental pillar. The social dimension and the citizen skills dimensions (through attempts to tackle ethical issues) are also contained in ESD objectives. What is notable is that, although some ESD objectives are explicitly or implicitly contained in the Biology courses, implementation strategies are left to the discretion of the teacher educators.

### *Swedish teacher educators' views*

In the interviews with the two participating teacher educators from the Swedish university, it was interesting to note that both participants were familiar with the SDGs, albeit admitting that they were not in a position to recite them, and understandably so. They both agreed with the fact that sustainable development and ESD were broad concept areas that included the interaction of ecological, social and economic dimensions. SWL2 feared that the concepts risked being everything and nothing. SWL1 expressed how he understood the concepts as follows:

> In general, SD and ESD are a way to address issues that include how humanity has to change its attitude towards the use of natural resources, not overexploit the resources, and sustainably use them.

According to theSWL1, humanity is interfering with nature's ability to recreate itself. He also thought that sustainable development is currently being seen from the bigger picture, such as water management, land resources and consumption, but he thinks questions should be asked about issues related to day-to-day choices and decision-making. He suggested that we should ask questions such as if we should drink milk, if we can replace something, and many other questions of this nature. It could be noted that the two teacher educators' perceptions of ESD were informed by the SDGs and the need to preserve nature and limit the current excessive use of natural resources. The environmental aspects of ESD were very prominent for the two teacher educators.

In terms of how SDGs informed the integration of sustainable development in the Biology classroom practices, the two Swedish participants seemed to have different opinions in terms of the influence of SDGs in their classroom practices. SWL1 seemed to believe that Physiology does not need to include sustainable development, and therefore the SDGs have very little influence since the course is about basic knowledge in physiology. He said:

> ... thus, sustainable development is not included in the Physiology course for student teachers. It is about basic knowledge that can then be used in a continuation of sustainable development education. But you need basic knowledge first.

This perception seemed to resonate with what SAL2 at the South African university indicated when she said that sometimes it is difficult to include ESD in

Biology content that focuses mostly on conceptual understanding. SWL1 also alluded to the fact that there was no time allocated for integrating SDGs in the (Physiology) course, so it was basically not done in his course.

SWL2 seemed to view ESD from sustainable development and SDG perspectives and therefore thought it was more applicable to Biology as a discipline than Biology didactics. Even though she thought ESD would be more applicable to Biology as a discipline, she observed that the integration was not quite deliberate nor planned for. She said:

> Although SDGs partly inform the integration of SD, but it is nothing that is planned for because there is a difference between Biology as a discipline and as subject matter in didactics [PCK]. The SDGs are more closely linked to the discipline [Biology], and there is no subject matter didactics [PCK] in the SDGs.

SWL2 highlighted that although she thought ESD would apply to Biology-content courses, she felt, nonetheless, that there were no policies that informed how the integration could be implemented. SWL2 also indicated that even if one tried to include ESD in the content courses, time would be a limiting factor.

However, because one Biology methods course included some ESD objectives, SWL2 thought there was room for integration. She said: "We (teachers in Science Didactics) integrate ESD and SD in a sub-course which is part of one of the Biology courses". Both participants agreed that if aspects of sustainable development emerge in other courses when students bring them up, then they could be discussed. They were quick to point out, though, that these instances are usually not purposely included during lesson planning. In the Swedish context, it could be noted that there was a sub-course that purposely focused on ESD. The sub-course, which constitutes one credit spread over one semester, is given to all student teachers, regardless of subject specialisation in the teacher education programme. This was not the case in the South African context. SWL2 considered the teaching of ESD to future Biology teachers to be very important. She suggested that student teachers could be made to work with authentic tasks and examples taken from real-life situations.

The Swedish teacher educators' perspectives on how the teaching of ESD to student teachers prepared them for their future classroom practices were further probed. SWL2 acknowledged that student teachers need to understand the complexity of the concept (sustainable development) and be presented with examples of how they could work with it in the teaching and learning situation. SWL1 believed that if there were enough time, it would have been beneficial to build a course that included ESD-application ideas and principles in the basic Physiology course. SWL2 lamented that there was a lack of practice by student teachers to integrate sustainable development in teaching during "their placement [teaching practice]". She explained that this practice would have enabled them to link their experiences to theory and to transform their knowledge to enable the teaching of ESD in their future classrooms. SWL2 said: "Student teachers should be receiving more teaching on working with ESD actively and not only lectures; however, but the lack of resources also contributes to these shortcomings". It was also important to find out from these participants if the teacher education was adequately preparing teachers to incorporate sustainable development in their classroom practices. SWL1 said he would not know since he does not incorporate sustainable development in his Physiology classes. SWL2 believed that the effectiveness of the ESD preparation was determined by the students' attitudes. She said: "It depends on how ambitious the students are. Those with interest and prepared to learn would benefit, but those that just want to complete the course would probably not gain that much knowledge". She also acknowledged that the sustainable development and ESD preparation received was inadequate. Her concern was that, since student teachers often teach as they have been taught, there is a risk that they may not be able to apply their knowledge about ESD in the classroom. She was very concerned that student teachers lacked practice in how to incorporate ESD into the school curriculum.

## Discussion

The purpose of this study was twofold. First, the study explored how policies guide future Biology teacher preparation for ESD at the two universities, one in South Africa and one in Sweden. Second, the study explored how Biology teacher educators prepared student teachers for ESD in these universities. In terms of how policies guided the preparation of student teachers to teach ESD, some notable findings were made. From the documents that were analysed, which included course outlines, it was noted that some Biology topics by default presented opportunities for the teaching of ESD. The most prominent topics fall under the broad area of ecology, making the environmental pillar of the ESD conceptual framework easier to address than others. The existence of topics in the documents that coincided with some of the SDGs (United Nations, 2018), such as numbers 2, 6, 12, 13, 14 and 15 (see Appendix), seemed to be an implicit way of suggesting that ESD should be included in teacher education However, in the case of South Africa, where the policy documents do not contain any broad or specific objectives to address ESD, only the topics related to the SDGs work to guide teacher educators in the preparation of Biology teachers for ESD.

What can be taken from the above finding is that the global sustainable development policies such as the SDGs and their predecessors also served to guide how student teachers may be prepared for ESD. All four participating teacher educators professed knowledge of the SDGs and some conceptions of sustainable development and were able to relate them to student teacher preparation for ESD. Whilst the above findings were true for the South African and Swedish contexts, the teacher education programme in Sweden, however, has a course that deals with ESD teaching. The existence of the course improves the opportunities to learn that are available to the student teachers. The sustainable development contents of the policy documents are more biased towards the environmental pillar. Social issues are also visible, but to a lesser extent.

The appearance of the social issues is in line with one model of integrating sustainable development in Science education, according to Eilks (2015) and Burmeister et al. (2012). However, the pedagogies for ESD, economic and citizen skills pillars were not explicitly included. It is our opinion that teacher educators with well-developed ESD agencies would be able to make the connections in their classroom practices.

The findings made when addressing the second research question on how teacher educators prepared Biology teachers for ESD are consistent with the findings from the first research question. It was noted that the participants' previous experiences and perceptions with ESD influenced how they engaged ESD in the classroom. Experiences were developed on previous work experiences in the area of sustainable development. Perceptions were developed both from the previous work experiences and the knowledge on sustainable development issues such as the decade for ESD, millennium goals and SDGs. From the previous experiences and perceptions of ESD, participants developed "an agency" which they use to prepare the student teachers. The participants' agencies in the South African case study seemed to have been influenced strongly by their previous work experiences related to sustainable development in addition to the knowledge of the millennium goals and the SDGs. The participants' agencies in the Swedish case study seemed to have been influenced by perceptions of sustainable development issues. Similar to the findings made from the document analysis, the participants' experiences and perceptions of ESD were biased towards the environmental and the social aspects of sustainable development.

Using the experiences and perceptions of sustainable development, the participants further expressed how they thought ESD should be taught to future Biology teachers. However, the perceptions they expressed on how student teachers should be prepared for ESD revealed some challenges and tensions that had to be negotiated. First, there was a feeling that the outcomes of some Biology-content modules, such as conceptual understanding, discouraged the teaching of ESD. If they considered including ESD, the limited time allocated for teaching still worked as a deterrent. Nevertheless, there was a feeling in both university contexts that some Biology-content topics presented better opportunities for teaching ESD than the methods courses. The outcomes of the Biology-content courses did not purposely include ESD to guide student teacher preparation. The Swedish teacher education programme, however, offers a sub-course that is specifically focused on ESD. The course presented enhances opportunities for ESD preparation in addition to other implicit ways. The classroom strategies that are used to teach ESD to student teachers heavily depend on discussions. It seemed the participants felt that more engaging methods could be used that allow the student teachers to actively participate in learning. The suggested methods included activities grounded in real-life issues and opportunities to teach ESD during teaching practice. Although the teaching of ESD to student teachers was mostly erratic, it was hoped that they would be grounded in sustainable development issues and be able to include them in their future classrooms. The participants felt that through their efforts, the student teachers could develop knowledge, values and ethics in the environmental, social, citizen skills and pedagogies dimensions of ESD. The economic pillar was neither explicitly nor implicitly mentioned, which suggests that it was an undervalued dimension in the preparation of Biology teachers for ESD. Tsakeni (2018) has also observed that the economic pillar of ESD was not promoted in the South African Physical Sciences school curriculum. The findings also seem to point to the erratic way in which ESD is included in the preparation of the Biology teachers in the two institutions studied. Specific outcomes for ESD were generally absent or implied, which compromises the manner of implementation. Implementation depends more on teacher educator agency than policy imperatives. However, Karaarslan and Teksöz (2016) call for more effective and comprehensive ways of preparing Science and Technology teachers for the implementation of ESD in their future classrooms.

## Conclusion and recommendations

The debate on sustainable development in the Anthropocene includes the role of education as one of the solution pathways. Teachers around the world are responsible for integrating sustainable development education (ESD) into the classroom. It requires teacher education programmes for Biology teachers to adapt to the requirement of sustainable development. This chapter addresses the challenges of how to prepare prospective Biology teachers to implement and integrate ESD in the classroom. In general, policies guide the preparation of Biology teachers for ESD in a very implicit way with a strong bias towards the environmental dimensions of ESD. The Swedish teacher education programme for secondary school teaching, however, contains a small course on ESD, which seemed to improve opportunities to learn. In the absence of explicit policy guidelines, the preparation of Biology teachers for ESD heavily depends on teacher educators' agencies, beliefs and perceptions developed from previous work and learning experiences in varied sustainable development areas. This study recommends that ESD should be systematically integrated across all the courses that student teachers take during their preparation, including teaching practice. This integration provides an opportunity for curriculum developers to engage in curriculum decolonisation processes aimed at knowledge generation by local players and framed around ESD. However, this integration should be well articulated in the policy documents, including course outlines, emphasising all pillars of ESD to provide the teacher educators with a clear direction. The use of the case study composed of a research site in the Global North and another from the Global South has revealed ESD integration in Biology teacher preparation remains understated and undervalued. The collaboration and exchange of experiences and data from the respective teacher education programmes have broadened our understanding of the universal nature of the challenges of realising ESD in the preparation of prospective Biology teachers.

The North and South cases make visible the varying and sometimes the lack of pedagogical content knowledge of contextualising and operationalising ESD within specific Biology courses. Further studies are recommended to explore how the economic pillar of ESD can be well represented in Biology and teacher preparation for ESD.

## Acknowledgements

The researchers in this study were supported by STINT and Thuthuka NRF South Africa.

## Appendix: Sustainable development goals (United Nations, 2018)

The 17 SDGs

1. End poverty in all its forms everywhere.
2. End hunger, achieve food security and improved nutrition and promote sustainable agriculture.
3. Ensure healthy lives and promote well-being for all at all ages.
4. Ensure inclusive and equitable quality education and promote lifelong learning opportunities for all.
5. Achieve gender equality and empower all women and girls.
6. Ensure availability and sustainable management of water and sanitation for all.
7. Ensure access to affordable, reliable, sustainable and modern energy for all.
8. Promote sustained, inclusive and sustainable economic growth, full and productive employment and decent work for all.
9. Build resilient infrastructure, promote inclusive and sustainable industrialisation and foster innovation.
10. Reduce inequality within and among countries.
11. Make cities and human settlements inclusive, safe, resilient and sustainable.
12. Ensure sustainable consumption and production patterns.
13. Take urgent action to combat climate change and its impacts (noting agreements made by the UNFCCC forum).
14. Conserve and sustainably use the oceans, seas and marine resources for sustainable development.
15. Protect, restore and promote sustainable use of terrestrial ecosystems, sustainably manage forests, combat desertification and halt and reverse land degradation and halt biodiversity loss.
16. Promote peaceful and inclusive societies for sustainable development, provide access to justice for all and build effective, accountable and inclusive institutions at all levels.
17. Strengthen the means of implementation and revitalise the global partnership for sustainable development.

## References


Abdullah, S. I. S. S., Halim, L., & Shahali, E. H. M. (2011). Integration of environmental knowledge across biology, physics and chemistry subjects at secondary school level in Malaysia. *Procedia – Social and Behavioural Sciences*, *15*, 1024–1028.

Borg, C., Gericke, N., Höglund, H. O., & Bergman, E. (2014). Subject- and experience-bound differences in teachers' conceptual understanding of sustainable development. *Environmental Education Research*, *20*(4), 526–551.

Burmeister, M., Rauch, F., & Eilks, I. (2012). Education for sustainable development (ESD) and chemistry education. *Chemistry Education Research and Practice*, *13*, 59–68.

Creswell, J. W. (2007). *Qualitative inquiry and research design: Choosing among five approaches* (2nd ed.). Thousand Oaks, CA: Sage.

Cuenca-García, E., Sánchez, A., & Navarro-Pabsdorf, M. (2019). Assessing the performance of the least developed countries in terms of the millennium development goals. *Evaluation and Program Planning*, *72*, 54–66.

Department of Education. (2011). *The national curriculum statement. Grades 10–12 (life sciences)*. Pretoria, South Africa: Government Printers.

Department of Higher Education and Training (DHET). (2011). *Minimum requirements for teacher education qualifications.* Pretoria, South Africa: Government Printers.

Eilks, I. (2015). Science education and education for sustainable development: Justifications, models, practices and perspectives. *Eurasia Journal of Mathematics, Science & Technology Education*, *11*(1), 149–158.

Grimm, Y. S. B. B. K. (2008). Achieving transformative sustainability learning: Engaging head, hands and heart. *International Journal of Sustainability in Higher Education*, *9*(1), 87–98.

Holbrook, J. (2009). Meeting challenges to sustainable development through science and technology education. *Science Education International*, *20*(1/2), 44–59.

Karaarslan, G., & Teksöz, G. (2016). Integrating sustainable development concept into science education program is not enough: We need competent science teachers for education for sustainable development – Turkish Experience. *International Journal of Environmental and Science Education*, *11*(15), 8403–8425.

Lewis, S., & Maslin, M. (2015). Defining the Anthropocene. *Nature*, 519, 171–180. https://doi.org/10.1038/nature14258

Muralikrishna, I. V., & Manickam, V. (2017). *Environmental management: Science and engineering for industry.* Oxford: Butterworth-Heinemann.

Robeyns, I. (2005). Selecting capabilities for quality of life measurement. *Social Indicators Research*, *74*, 191–215.

Schild, R. (2016). Environmental citizenship: What can political theory contribute to environmental education practice? *The Journal of Environmental Education*, *47*(1), 19–34.

Stanisic, J., & Maksic, S. (2014). Environmental education in Serbian primary schools: Challenges and changes in curriculum pedagogy and teacher training. *The Journal of Environmental Education*, *45*(2), 118–131.

Tsakeni, M. (2018). Opportunities for teaching sustainable development through the chemistry component of CAPS physical sciences. *African Research Journal in Mathematics, Science and Technology Education*, *22*(1), 125–136.

UNECE. (2011). *Learning for the future: Competences in education for sustainable development.* Retrieved July 15, 2019, from http://www.unece.org.unecedev.colo.iway.ch/fileadmin/DAM/env/esd/01_Typo3site/ExpertGr oup Competences.pdf UNESCO

UNESCO. (1997). *Thessaloniki declaration*. Gland: UNESCO.

United Nations. (2015). *Transforming our world by 2030: A new agenda for global action Zero. Draft of the outcome document for the UN Summit to adopt the Post-2015 Development Agenda*. New York, NY: United Nations Publications.

United Nations. (2018). *The sustainable development goals report 2018*. New York, NY: United Nations Publications.

Woodridge, M. and Zimmermann, M. (2015). From MDGs to SDGs: What are the sustainable development goals? ICLEI briefing sheet. *Urban Issues*, *1*, 1–4.

# 8 Culturally embodied learning as instructional practices in science education

## Different contexts and different meanings

*Emmanuel Mushayikwa, Miranda Rocksén, Sakyiwaa Danso Boateng, Samuel Khoza and Git Börjesson*

## Introduction

The vision of the United Nations (UN) Decade of Education for Sustainable Development (UNESCO, 2021) was "of a world in which everyone has the opportunity to benefit from education and learn the values, behaviours and lifestyles required for a sustainable future and for positive societal transformation" (Wals, 2009, p. 8). Almost twenty years after its institution, today this seems equally relevant and demanding. Yet, understanding education's role in sustainable development in Africa, means not to solely fit this into the hegemonic education discourse based on Western epistemology. Through research about indigenous knowledge, indigenous knowledge systems and indigenous science, we learn from indigenous cultures across the globe (i.e. Snively & Corsiglia, 2001; Zidny, Sjöström, & Eilks, 2020), not least in Africa (Shizha, 2017). At the same time, in many parts of the world, communities articulating distrust in scientific knowledge challenge the educational endeavour of shaping the future world, in what have been called the post-truth era (Feinstein & Waddington, 2020). If the global discourse on education for sustainable development (ESD) incorporates aspects of indigenous knowledge systems, this could be one way to open up ground-breaking avenues in the quest for a more sustainable future (Breidlid, 2009), but it would not suffice. In order to deepen understandings and be able to discuss sustainability issues in a broader, culturally embedded perspective, knowledge about situated science education practices is needed. However, there is a gap in the literature with limited data available on how to understand teaching practices in different cultural settings for example with regard to provided learning opportunities from holistic and embodied perspectives.

ESD is defined as "an education that allows every human being to acquire the knowledge, skills, attitudes and values necessary to shape a sustainable future"

DOI: 10.4324/9781003190158-11

(Leicht, Heiss, & Byun, 2018, p. 35). According to Leicht et al. (2018), ESD aims to develop competencies that enable individuals to reflect on their actions by taking into account their current and future social, cultural, economic and environmental impacts from both a local and a global perspective. According to UNESCO (2017), ESD also relates to the core of teaching and learning, and integration of sustainability concepts into the curricula, to achieve sustainability-related learning outcomes. Here, teachers are seen as the most critical drivers for restructuring educational processes and moving educational institutions towards sustainability. To ensure the smooth operation of ESD, teachers must develop sustainability key competencies through innovative teaching and learning instruments, although the role of science education may be contested. Even if science education plays an essential role in the socio-economic development of a nation and should serve humanity by providing youth with knowledge of society, nature, life and sustainable environment, there are factors militating against broad access to scientific information and the teaching and learning of science in the south as well as in the north.

In Africa, evidence emerging from empirical studies suggests that culture is one of the militating factors against the teaching and learning of science (Anamuah-Mensah, 1998; Okere & Keraro, 2002; Shumba, 1995). Cobern (1993) argues that in developing countries, there is a need to ask "questions about worldview and the compatibility of various non-Western worldviews with modern science" (p. 2). Thinking through the assertions of Odhiambo (1972) and Ogunniyi (1988) on modern science as a product of a Western culture, this may not be readily amenable to the African worldview. Learners come into the science classroom with their own prior knowledge and experiences. For these learners to revisit their current thinking, they need to link up new information and create inferences and relationships between old insights and new ideas. Drawing on Simayi and Webb (2019) and their notion on knowledge restrictions when teaching some topics in sciences, Shumba (1995) elucidates the conflicts that arise in the science classroom because some science topics might be taboo to specific cultures. For instance, in some Xhosa communities in the Eastern Cape Province of South Africa, it is deemed a taboo to use standard human reproductive terms to teach human reproduction in Biology. Indeed, this position exists in many African societies. Okoye and Okeke (2007) argue that it has been difficult for Western scholars to clarify several natural occurrences and issues in African life using scientific knowledge. This has resulted in a divergence between learners' daily adventures in the home and the science classroom with most of the learners having significant difficulties in describing, understanding and predicting natural phenomena using Western scientific ideologies (Engelhardt, Corpuz, Ozimek, & Rebello, 2003).

In this chapter we have a methodological focus, presenting and analysing data from interventions into teaching practices. The step between education for a sustainable future and culturally situated science education is not far if we are to develop a sustainable and quality education for all. Our approach is based on two assumptions. Firstly, we assume that science teaching can be improved by considering sociocultural and holistic approaches, of which embodied learning experiences is seen as one aspect (Johnson-Glenberg, Birchfield, Tolentino, &

Koziupa, 2013). In acknowledging the experience of the body in pedagogy, we join a growing number of other scholars such as Andersson and Risberg (2018), Johnson-Glenberg et al. (2013), Munro (2018), Stolz (2015) and Sund, Quennerstedt and Öhman (2019). These scholars are offering design pedagogies to be considered within pedagogical spaces as an essential element of practice. Secondly, we assume that multicultural science education, including successful integration of science and indigenous knowledge systems in science education, supports sustainability, appropriate internalisation of values and modification of students behaviour (Okoye & Okeke, 2007). While emphasising culture as an essential aspect following Atwater (1994), we acknowledge that several factors influence a child's conceptual development (Ogunniyi, 1985). Before a child starts school, he/she has a repertoire of cultural beliefs, myths and explanations as well as everyday life experiences of natural phenomena and has further been exposed to and partaken in several cultural practices. Some of the cultural practices may be at variance with the conventional or traditional culture of schooling, and with the way scientific concepts are used to explain phenomena in scientific reasoning. The purpose here is to explore the concept of culturally embodied learning as an interpretive frame for the analysis of two teaching interventions and their impact on learners and teachers experiences towards teaching and learning in physical sciences. Although we are "insider researchers" to our respective national contexts, we are at the same time "outsider researchers" to each other' contexts (Erickson, 2010). Hence, being aware of the limitations with regard to generalisability this entails, here we take the route approaching the subjectivity of teachers and learners, and the context dependence of meaning (Englander, 2019).

## Theoretical and methodological frame: Culturally embodied learning

One classical dualism documented in western philosophy at least during 500 years since René Descartes in the 17th century is the one between body and mind. It is a dualism that still seems to balk the understandings of learning and teaching in many settings. In several school sceneries, the body is often put in the background with regard to studies of teaching, learning and pedagogy in the classroom and is instead predominantly dominated by discussions of mental functions and processes (i.e. Hager & Johnsson, 2012; Ord & Nuttall, 2016). With sociocultural theories on learning on the other hand the social and material ties that form the individual in his or her environment are taken into account, including culturally embedded and embodied knowledge (e.g. Säljö, 2009). In the words of Odora Hoppers (2002) culture is the collective property of a group and manifests itself in learned behaviours, forming a pattern that shapes values from generation to generation. Seen as a product of human activity, culture may include: instructional practices; educational institutions and discipline; educators; the functioning of the educational institutions as well as intercultural understanding among others. Embodied learning experiences are embedded in culture in which inherent, unconscious societal values lead to preferences over certain things and loyalty to the group (Chilisa, 2012). From the view of the individual learner, embodied learning means coming to know oneself better and the world around, neither as an abstract object nor as an instrument, but as a "lived body" that senses and does the sensing in a meaningful way (Stolz, 2015). Drawing from Merleau-Ponty (1945/1989)'s work, Thorburn (2008) and Stolz (2015) reconceptualized the nature of embodied learning experiences and their implications to education. Merleau-Ponty contends that the totality of the whole "body-subject" provides a devising idea relating to the body and the environment. Borrowing heavily from Munro (2018), we defined teaching with the role of embodied learning in mind as "the deliberate use and recognition of multimodal body-mind activities and strategies to facilitate shifts in perspectives, perceptions, paradigms, behaviour and actions" (p. 7). With these writings as an influence and by using the concept *culturally embodied learning* as a theoretical frame in this chapter, we approach learning as embodied and embedded in culture.

This research was positioned in the interpretivist paradigm, because it is characterised by a need to understand the world as it is from a subjective point of view and seeks an explanation within the frame of reference of the participants (Ponelis, 2015). In addressing this problem, we were informed by Stake (2013)'s multiple case study approach which foregrounds the collection of rich data, with attention to "thick description", "experiential understanding" and "multiple realities" (p. 43). Stake focuses upon the analysis of patterns and differences within and across data sources, from which themes and findings emerge: confidence in particular findings is developed through triangulation of multiple perspectives.

The study investigates two interventions that were conducted separate from each other in South Africa and Sweden, two countries with different cultural and school demographic backgrounds and teacher-learner characteristics. According to Yin (2015), mixing cases with other research methods enable researchers to address broader and more complicated research questions than the single case study alone. The intervention in South African was located to the Eastern Cape province which is one of the nine provinces in the country. South Africa has 11 official languages that are spoken with Sign language being the additional one to make it to 12. The Eastern Cape is known for its rurality landscape with isiXhosa as the main spoken language. The province runs along the Indian ocean with farming as one of their main daily duties. Learners and teachers in the provinces' schools are so immersed in indigenous and cultural activities and this already defines how they learn in formal platforms like schools. They still engage in traditional criminal courts that are supervised by a village chief where they would sit around the fire discussing issues over home brewed traditional beers and smoking pipe. The intervention in Sweden was smaller. It was conducted as a project within primary teacher education, involving the two co-authors as participants. In Sweden play is important in the primary school, and it is emphasised in curricula. The syllabus for natural sciences mentions for example that teaching the area force and motion is preferably done in the playground. In this chapter, we understand the playground as a representation of a children's culture, of which the vast majority of student teachers in Sweden have their own previous experiences from playing in playgrounds in their childhood. Both interventions aimed at incorporating embodied learning in instructional practices and to make science relevant to both the learner and his/her environment. The researchers sought to understand how culturally embodied learning played out in the classrooms where it was actually practiced and so the selection was biased towards teachers who showed evidence of willingness to use this instructional strategy in their classrooms (Creswell, 2009). The interventions were however developed separate from each other with the joint theoretical frame added post the data collection, in an abductive design. In the next two sections the interventions, with data and methods of data collection, are presented, beginning with the intervention in South Africa. In an attempt to bring to life something of the two contexts to the reader, contexts are presented first, followed by some emerging themes which are presented through data examples and analytical comments.

## The South African context

Participation in the study was both voluntary and purposive in that only those teachers who showed consistent commitment to implement the culturally embodied learning instructions as a teaching strategy in their classrooms were approached. A total of four teachers and 200 learners in their intact classrooms were selected on voluntary basis. All the classes were mixed gender. Abilities differed within and across classes. Four teachers were selected for the study because they were keen to explore the pedagogy in their respective classrooms. The selected schools were initially visited to meet with the schools' principals and physical science teachers to facilitate the administration of the instruments. Data was collected during the Term 3 of the academic year, after obtaining ethical clearance from the province and the principal. The study was carried out in four phases. In the first phase, four teachers were trained to use culturally embodied learning tools to teach lessons in electricity and magnetism. In the second phase of the research, non-participatory classroom observation tool was used to observe four (n = 4) physical science teachers' use of culturally embodied learning instructions that supported learners' learning and learners' cultural backgrounds. In the third phase, semi-structured interviews were conducted on a sample of 16 learners purposively selected. After the classroom observations and learner interviews, the last stage of the data collection process included semi-structured interviews of the four participating teachers. They were interviewed for the purposes of corroborating responses and to get a deeper understanding of teachers' perceptions and conceptions of culturally embodied learning.

## Learners and teachers in interviews about culturally embodied learning

The four teachers willingly agreed to re-sequence their pacesetter content to teach electricity and magnetism in Grade 11 during the period of the study, and the 16 interviewed learners took part. During the interviews with teachers and learners four significant themes emerged. Each theme represents a critical feature in relation to this case of implementing culturally embodied learning in a physical sciences classroom. These themes concerned (1) how learners' home backgrounds influence their learning, (2) how learners' home backgrounds influence their relation to the teacher, (3) teachers' perspectives on scientific knowledge as socially and culturally influenced and (4) teachers' views on using culture as a lens to change teaching learning processes in the science classroom.

The first theme was about how learners' backgrounds influence their learning and did include discussions about the influence from home cultural practices on school and particularly learning in the physical sciences, as well as some differences between learning at school and at home. Responses from the learners indicated that they relied heavily on verbal information received from adults in authority, without questioning, basically in the form of directives. This could be attributed to the socialisation of the learners in which respect for authority is seen as paramount. This could be influenced by the set-up that the learners grow under where authority is of paramount importance which emanates from a traditional setting led by an elder (chief) in the village. Of interest are what Tsepo and Aviwe expressed (names are pseudonyms):

*Tsepo:* I learn the most at home by listening and following orders from an older person than me.

*Aviwe:* I learn from my parents or an adult in authority mostly through verbal information when they talk to me or confront me to do the right thing.

As in these two examples learners' responses indicated that they mostly learn from verbal information in the form of directives. References were also made by the learners on learning from visual information, mostly through observation. In one example, a learner narrated that, when an elderly person wants to demonstrate how a particular food is prepared, the child is asked to look at how it is done after which the child is made to try it out on another day until she becomes perfect in preparing that particular meal:

*Avela:* I learn things when I see them doing it as well. My parents show me how certain things are done.

In the interviews the learners pointed towards that at home emphasis is on taking instruction and learning to take part in a practice by processing visual and verbal information in learning situations. Learners' responses indicated overwhelmingly differences between home learning and school learning.

*Okuhle:* At home, I am more likely to remember what I learnt from closer relations, whilst in school, you are forced to learn even when you are not willing to.

This is an indication that learners come to physical sciences classroom with misconceptions arising from cultural beliefs that they hold. Several learners commented on the fact that their cultural practices have some influence on their learning at school. Here is Iviwe:

*Iviwe:* In my culture, I am expected to respect adults and not to talk back to any adult in authority. In school too, I have been told to respect all my teachers and listen to what they say and ask me to do.

Other research (e.g. Breytenbach, 2013) has indicated that culture influences behaviour and therefore it would seem likely that differences in the cultural dimensions of groups of people would lead to differences in their approaches to working and studying. What the learners in this study express about physical sciences is in consonant with Ododo (2014) who opined that learners bring many of their cultural practices they have heard or held into biology classes. According to Yip (2001), personal experience and opinion are related to cultural factors prevalent in the society. This is because cultural position at home gives room for conflicting traditional beliefs and practices which differ from situations at school. A common trend through the learners' responses was that they mostly apply their cultural practices in the physical science classroom. One learner narrated:

*Avela:* I apply some of my cultural practices at home in school, for example, when talking to an adult do not put your hand in your pocket. Respect for other people beliefs and practices, and at school, there is a common practice which we all subject to.

Learners' responses corroborate the study by Samba (2003) who found that learners cannot learn new concepts if alternative models that give explanation about a phenomenon exist in their mind except a platform is made available to them to confront their misconceptions reconstruct and internalise their knowledge. In addition, Okebukola (2002) orated that learners who have had themselves established in their cultural belief are likely to find the study of science mystifying because of the likely conflict between their anthropomorphic view of their world and the mechanistic views presented in science. This implies that any mistaken belief from cultural background brought into physical science classroom by learners would affect their learning in the subject.

The second theme illuminated that learners' home backgrounds influence their relation to the teacher. The learners emphasised the need to respect the teacher as the one in authority and not to question his authority. Learners look up to their teachers as having the last word in everything they teach. One learner narrated:

*Sinalo:* Sometimes, I choose to keep quiet if my view point is not considered by the teacher to be valid. This is because; I was raised at home not to engage in any form of argument with teachers or elderly people.

What this pupil says is an indication of one pupil's personal experience when participating in the science classroom. The introduction of other opportunities for participation and discussion with the teacher, might have positive effects on this and other pupils' learning. If teachers learn to embrace their learners' cultural backgrounds in the teaching-learning processes and become flexible in terms of modifying teaching strategies in their science classrooms, this could also improve learners' performance in the subject. With this in mind, and as indicated by Ingalls, Hammond, Dupoux and Baeza (2006), the educational goals could be enriched and could also alleviate fears that relate to loss of group identity, unfriendliness and uneasiness during the teaching learning processes. However, the foundation of a healthy relationship is built as teachers take the opportunity to interact and learn from their learners. The way learners may react to adult in authority in their culture may be different with what is expected at school. For example, keeping away eye-contact might seem inappropriate in the learners' home culture, whereas in school, it is acceptable for a learner to keep eye-contact with the teacher. One learner commented:

*Samani:* The relationship I have with my teacher is just like the relationship I have with my parents. Since I was brought up to keep quiet if unhappy and not to stand up to my teacher, I don't complain about my teacher to anyone.

However, in this case, there is not much arguing of one's point of view even when one feels unhappy about some issues in class lest they may be mistaken for being disrespectful.

The third emerging theme was about teachers' perspectives on scientific knowledge as socially and culturally influenced. Some teachers were of the view that science is part of social and cultural traditions:

*Miss Cassandra:* In my view, I think science is everywhere and therefore, ideas of science are affected by the cultural, social and political settings of the society and would determine when and how science should be conducted in that particular environment.

An example to illustrate the perspective of science and culture reciprocally effecting each other could be when a community is affected by a pandemic such as COVID-19. Indigenous cultures generally attribute religious causes to pandemics and chronic diseases. While scientists generally attribute the same to natural causes which can be overcome without divine intervention.

In another interview, Mr Kelvin compared the scientifically based knowledge about human albinism, which is caused by a mutation in genes controlling melanin production, with the belief in his culture that this is the result from the mother being mean during pregnancy:

*Mr Kelvin:* My cultural and societal values and ideologies are often proven wrong by scientific knowledge. For example, in my culture, there is a belief that albinism is caused by a spell which was evoked on the mother as a result of her being mean to other people during pregnancy. In contrast, science explains albinism as a lack of a skin pigment due to mutations. This is evidence that scientific knowledge is not socially and culturally influenced.

The two perspectives on knowledge, seen as socially and culturally influenced and seen as separate from each other, that appear in the interviews with teachers illuminate a tension existing not only between teachers and their pupils and families, but also between teachers. However, science as an endeavour and phenomenon is not conceived and operated in a cultural and environmental blankness (Inokoba, Adebowale, & Perepreghabofa, 2010). Rather, science is a social phenomenon which is greatly influenced by the prevailing cultural practices and worldviews of mainly Western social cultural groups. Cultural values and expectations of these societies determine what and how science is conducted, interpreted and accepted.

The fourth theme concerned teachers' views on using culture as a lens to change teaching learning processes in the science classroom. A common trend identified in the teachers' responses was the idea of involving cultural activities in teaching, using scenarios applicable to learners' culture in the teaching of science. Two teachers narrated:

*Miss Merlene*: I think understanding the diversity of the classroom is the first step in laying the foundation of teaching science, and further providing appropriate culturally based instructions that takes into account learners' prior preparations, and incorporating this in familiar hands-on activities that engage learners in the teaching learning process.

*Miss Cassandra*: I have a belief that if teachers consider the differences learners bring to the science classroom and provide instructions that incorporate the embodied learning together with practical oriented instructions will go a long way to engage learners in the teaching learning process for sustainability.

These teachers also recognised the importance of considering learners cultural background in the teaching learning process. Mr Kelvin narrated:

*Mr Kelvin:* In most instances, I relate my teaching to their cultural background, making comparisons here and there giving examples related to their home culture. Sometimes too, I explain concepts in Physical sciences in a form of a story, something they can relate to, then I use the same scenario to relate to the concept that I am teaching about in the classroom, considering their learning style preferences. Embodied learning in a context of their cultural background in my physics classroom.

In the interviews the teachers expressed positive attitudes towards the idea of doing investigations related to the culture of the learners to familiarise oneself with the instructional practices that would be much convenient to the teacher and the learners at large. Some teachers expressed the view that embodied learning in the science classroom will facilitate smooth transfer of knowledge:

*Mr Albert:* As a teacher, it deemed right to know learners cultural background and embrace their culture and take preferences in the way they behave, particularly the way they learn and infuse it in the teaching learning processes through embodied learning activities.

*Miss Cassandra:* Teachers' should become flexible to alter their instructional strategies in their classrooms by considering learners needs and cultural identities through embodies learning activities.

Differentiated instructions that relate to teachers' use of culturally congruent content and instructional methods that reflect the unique cultural, scientific and contextual perspectives of the students they teach were some of the views expressed by the respondents in South Africa.

## The Swedish context

In the Swedish case, the intervention including embodied learning experiences was conducted in relation to a module about force and motion in primary teacher education. The module was a regular part of a mandatory 10-week course on the teaching and learning of science and technology in the education of primary teachers and included one two-hour lecture and one three-hour workshop performed in groups of 25–30 students. The topic *Force and motion* is one among the core contents for Grades K-3 in the Swedish curricula. In a commentary material it is explicitly stated that the teaching of this content ought to allow children experiences of play and movements (Swedish National Agency of Education, 2017). It is mentioned that through these kinds of experiences, children can make observations of *force and motion*, *gravity, friction, balance, centre of gravity* or *equilibrium*. In order to prepare teacher students for organising teaching and learning activities in the playground in their future professional life, in this intervention the idea was to provide them with experiences, and expose them to situations, similar to what is described as teaching and learning activities in school, at the same time as we wanted them to gain a deeper understanding of physics concepts. Important in the planning of the workshop activities was a web resource about the playground as a physics laboratory, provided by National Resource Center for Physics Education (http://www.fysik.org/english/). Here, we want to illuminate the contextual aspects involved in the workshop that we identified as affordances and obstacles in order to reach the teaching and learning goals. The dataset consisted of the developed teaching and learning materials, personal teacher plans and course documentation. The analysis of the workshop focused on time, body/space and relationality (after van Manen in Stolz, 2015) and resulted in three themes presented as three vignettes together with figures, followed by analytical commentaries.

## Narration of a teacher educator

**Vignette 1: The workshop as embodied activity**

*We packed a trolley with equipment such as soft toys, slinky-springs, balls of different sizes and weight, literature and chalk for drawing on the asphalt (Figure 8.1). A couple of weeks in advance we visited a playground ourselves to practice and prepared written instructional material, which we also laminated, in order for it to sustain in bad weather. The instructions to students described a number of different investigations they were supposed to perform at the playground, such as by the swing,*

*Figure 8.1* One of the authors prepared for the workshop

Photo by Git Börjesson.

*by the see-saw, at the slide or throwing and catching a ball. These investigations were designed as minor inquiries and inspired by the course literature. During each investigation the students were instructed to discuss and respond to formulated questions connecting the activity to physics concepts. On a freezing but sunny November morning we collected all the material and went to a big playground called Plikta, in the centre of Gothenburg, where we met the group of student teachers. Time flied, and after several hours of intensive work instructing and supporting the student teachers in conducting their inquiries, we packed our equipment into the trolley and returned to the university.*

### Vignette 2: Playing in the playground and the object of physics

*One of our teaching and learning goals involved to get the teacher students' to use physics concepts for explaining phenomena. In the slide, the task was to try different materials, such as cotton fabric, nylon fabric and so on, to sit on when gliding down. During this activity they were supposed to focus on using the concepts friction, velocity, angle, speed, gravity and surfaces in explanations of different sliding experiences. In the swing, the task was to take a seat with a soft toy with long arms and legs on the side (Figure 8.2). During this activity they were supposed to focus on the movement of the swing, identify positions of high and low velocity and*

*Figure 8.2* One of the authors moving in two directions in the swing

Photo by Miranda Rocksén.

*explain the movements of the soft toy. As long as they observed, described, experienced and perceived by their senses during their investigations, the activities in the playground went on smoothly. However, when they were challenged to explain the phenomena under investigation, using physics concepts, many were frustrated and lost their concentration. The complex environment including, sounds, other visitors, cold weather and many impressions made communication and concentration more difficult. At this point – the playground itself worked against our teaching and learning goals.*

**Vignette 3: Perspectives on learning opportunities**

*We found that one set of activities was a bit more comprehensible and enabled several of the student teachers to shift perspective, from learners to prospective teachers, favouring discussions about school pupils' learning opportunities. These were the activities connected to the seesaw, with the concepts balance, equilibrium and centre of gravity. The students were instructed to investigate their own balance, when moving the body and shifting their centre of gravity. They were supposed to investigate how the distance and mass effected the equilibrium between two sides of the seesaw. In these discussions, the teacher students initiated their own discussions about teaching matters. A recurring discussion was if a child with obesity would feel uncomfortable participating in the investigation on equilibrium at the seesaw. They discussed what this might mean for the pupil's learning opportunities and what precautions that could be made to avoid such situations to occur.*

Although we aimed at incorporating embodied learning in instructional practices, and to make science relevant to both the learner and his/her environment, we learned that the playground, as an environment for learning about physics, in fact is very complex. You can try out friction when you ride the slide and you can feel the acceleration in a swing, but the movement of a swing is complicated to explain and it was a challenge to do this in the playground environment. In general, it was difficult to discuss the details of the physics involved in the tasks during the workshop. For the student teachers, the combination of play and inquiry meant a particular challenge. In order to make meaning of the activities to be able to observe phenomena, and describe and explain them using the associated physics concepts, the teacher students needed more of our support than we had expected. Sliding was apparently more connected to talk about friction and motion whereas catching or throwing a ball was more into gravity and forces. Investigating changes in velocity, like in the swing and the throwing of a ball, was more difficult compared to the investigations on balance and centre of gravity. When you are sitting in a swing (Figure 8.1), your feet no longer reach the ground. Instead your weight rests on the material of the swing. The two chains that hold the rubber tyre are connected to a rack, which creates stability and mobility. Just taking the seat in the swing, will be enough to make it move. By conducting purposeful shifts of the centre of gravity of your own body, leaning forward and backward, these small and irregular movements result in bigger swings which are more regular, with similarities to that of a swinging pendulum and possibilities to increase the height of each pendulum movement. In Figure 8.2, the white arrows represent the direction of the movement of the object. The swing has just turned its direction and is now moving to the right (right arrow). The soft toy, the slinky spring and the legs, on the other hand, have not yet turned their direction. They are still moving to the left (left arrows). A moment later the swing has moved further and forced these objects to change direction. What is observed can be described using Newton's first law of motion: an object continues to move in the same speed and direction until acted upon by a force. In practice, the turn of the swing and the change in direction happen so quickly and are therefore not easy to observe. Documenting the playground activities on video, and an indoor summary of the experiences and the physics involved, could be a complement for achieving the teaching and learning objectives.

## Discussion

This chapter relates to the lessons we can learn from our exploration of culturally embodied learning as an interpretive frame for the analysis of two teaching interventions; the teaching and learning of science in high schools in Mthatha, South Africa and a small intervention in primary teacher education in Sweden. The reporting from interviews and autobiographic participant observations are performed in the settings of two different cultures and make it evident that the ways learners of the two countries learn vary, as well as that the work of a science teacher varies. Firstly, the South African context happened in the classrooms to learners who are from the same province of the Eastern Cape where isiXhosa is their mother tongue. These learners grew up under certain rules and context, which are different compared to when they are in a formal setting like a classroom. In the interviews with the learners and teachers in South Africa, we have shown how they reflect on the influence from learning at home for the learning in physics at school. On the other hand, the Swedish student teachers had their workshop in a local playground in order to be better prepared for organising teaching and learning activities in the playground in their future professional life, as promoted by the national curricula.

## Embodied learning opportunities

In the Swedish setting, the students were given investigative tasks such as the Seesaw, the swing, the slide, throwing and or catching the ball to perform, these tasks seemed simple to the students because one could say they are part of their previous daily practices as children. It therefore became difficult for the students when the body did not constitute a means to knowledge but it reflected the students' reaction to the environment (Smyrnaiou, Sotiriou, Georgakopoulou, & Papadopoulou, 2016). Following Merleau-Ponty (1989), perception is anonymously presented to us by our senses, and when going by the slide, sitting in the swing and throwing a ball, for most of us these are unreflected motorised embodied habits. According to Sullivan (2018), both human-centred and technology-centred or technical-centred methods do have their respective benefits for student outcomes. Smyrnaiou et al. (2016) add that

> In embodied learning, new knowledge is affected by the conditions it is used and by the types of activities the student is expected to participate
> (Smyrnaiou et al., 2016, p 2).

The fact that the Swedish activities happened in the playgrounds, it is evident that the students' frustration could not only be related to new knowledge or the challenges that the activities came with, but also by the environment itself, because there were other people in the playground and the weather too could have had an impact to the students' learning. However, we notice that part of the teaching and learning activities performed at the playground was taking into account the everyday experiences of children in play, the activities that the students were exposed to were not new to them. Most students, if not all, are used to a swing, seesaw and throwing and or catching objects, but their previous experiences differed and the experiences in the workshop were affected by contextual factors (the playground). Even though active engagement and increased quality of learning was expected (Sullivan, 2018) this was not a clear-cut result. In the phenomenology of Merleau-Ponty (1989), the spatiality of one's own body is "always tacitly understood" (p. 87). It is by analysing one's own movement, that it can be better understood. In Figure 8.1, one of the chapter authors stands in an upright position. The body rests on the feet where the weight meets the solid ground. The left arm presses the soft fabric and fill of a large soft-toy towards the side of the body. The right hand holds the grip of the wheeled bag, which is filled with books and other equipment. The body is at a momentarily rest, a stop in the movement before the turning around and walking away. It is the movement of a teacher educator oriented towards conducting a workshop with student teachers. The movement, grip and holding of materials and objects is performed as if they were habitual. In the playground the teacher educator moves between groups of students involved in the different activities. It is by moving the own body in the swing, or by kneeling on the asphalt and drawing the movement of a ball throwing, that the teacher educator illustrates procedural and conceptual aspects of the different activities.

In the SA context, learning was a bit different because of the set-up that the learners from the Eastern Cape raised. These learners are taught to take instructions from the elders and not question them. New knowledge is affected by the conditions which it is applied to and the type of activities students are expected to participate in. This is attested by one learner from SA, Okuhle who said: "At home, I am more likely to remember what I learnt from closer relations, whilst in school, you are forced to learn even when you are not willing to". This adds to what Okebukola (2002) orated that learners who have had themselves established in their cultural belief are likely to find the study of science mystifying because of the likely conflict between their anthropomorphic view of their world and the mechanistic views presented in science. Perhaps this is also what the Swedish students in the playground went through. They did activities that were not new to them but their emotions showed a different expression and had an effect on the learning experience. Smyrnaiou et al. (2016), consider embodied learning as involving coordinated body movements for the achievement of learning goals and the development of affective connections to what is learned. A learning goal to be achieved combined with their emotional engagement. They contend that when planning an activity, the following should be considered: cognitive involvement to the topic, cognitive processes, representation of a scientific notion, body movements, expression of the student's feelings, clarity of instructions, holistic design of activities, student cooperation as well as the ability of students to apply acquired knowledge to new environments. Embodied learning conforms to holistic educational practice as it educates the whole person. In the Swedish intervention, our conclusion is that physics was a limited part of the embodied learning experience. Here, the teacher educators could have made more explicit to themselves and to the teacher students, what it meant for them to investigate and what they experienced with the body with the well-known physics concepts as tools. Compared to this, the student teachers by making inquiries in physics in the playground had to apply new procedural and conceptual aspects to already well-known habits of play.

## Culturally embedded views on teachers and learners and science

The findings illuminate the foundational influence of culture in the view on authority, the teacher, the learner, the subject. This then puts more demand on the teachers who believe that cultural and societal values and ideologies are often proven wrong by scientific knowledge. This was what one of the teachers based in SA responded when asked whether scientific knowledge is being socially and culturally influenced. Using the playground as the physical location of the teaching implied a movement from the ordinary classroom environment. However, even if the activities in theory are similar, the embodied learning experiences of being a child or an adult in the playground make them different activities in practice. It is relevant to ask what the different roles of being Pupil or Teacher afford for the learning experiences in a playground? In teacher education, we conclude the importance of discussing how the differences between being a student teacher – and a teacher educator – can be accounted for, in particular when planning for culturally embodied learning experiences? The starting point is – as with teaching and learning generally – the knowledge and experience of the learner. It is complex to foresee what can be expected learning opportunities because embodied learning experiences are subjective and often unreflected.

Teachers from the two cases were clear on issues of culture and how science can be linked to the learners' backgrounds. However, the two countries are different and culture and backgrounds will always be a serious and a contentious issue. The two countries can however use embodied learning but the suitability of it will not be the same. Science on the other hand will remain as rigid no matter the background. Although the two areas of physics represented here, electricity and magnetism, and force and motion, remain the same but as two distinct areas, the suitability for embodied learning experiences, representations or simulations may not be the same. While the SA activity was in the classroom, it seemed unfair to compare their classroom activities and that of the Swedish, which were held in the playground. The study showed that the two countries differed in terms of culture and learner backgrounds. However, these differences notwithstanding, teachers in both countries experienced similar challenges in trying to implement culturally embodied learning in the science classroom.

## Towards a global discourse in science teacher education

The chapter provides ideas for new features and benefits on the design and implementing of culturally embodied learning in the science classroom, with the overarching goal to contribute to systematic development of science education practices. In concluding we point to some specific implications of this study. The comparative cases illustrate the possibility of integrating specific culture with Western science and thus afford communities opportunities to adopt scientific reasoning into their cultural practices. Such an approach provides us with a prospect to introduce alternatives and developing knowledge beyond traditional empiricism and intellectual narratives. This could lead to the adoption of science-based solutions to environmental problems, and sustainable development initiatives.

The text in this chapter has evolved over time. As a group of authors we brought our personal and professional knowledges and experiences about science teaching practices as well as about the culture and background of the learners' in our respective countries. The two described interventions emerged as potential study objects along the way. By working with the text together, and taking part of the details of each other's cases, we have realised the immense global challenge involved in order to by 2030 ensure inclusive and equitable quality education and promote lifelong learning opportunities for all (SDG 4), but also that we have a lot in common. In order to build a foundation that meets and considers the learner, and his/her cultural background, we argue for international collaborations developing science teacher education, that embraces broad and inclusive approaches to teaching methodologies.

The findings of this study provide implications to curriculum developers and instructors in consideration of adjusting teaching modes to facilitate more effective learning using culturally based instructional strategies for the transmission of instructional objectives for learners coming from different cultural backgrounds. With better understanding of learning as embodied in culture, it is more likely that teachers can address group differences and apply different strategies in fulfilling the needs of learners' preferences to help in strengthening their potential competences and remedying possible deficiencies that may restrict their academic achievement in physical sciences.

The results of this study pave the way for further studies on the potential of culturally relevant science teaching analytic research frameworks and the design of culturally embodied science simulations for learning. Results from future studies focusing on relating and describing the process of embodied learning in science education could inform teacher education practices in order to combine the conceptual and more concrete forms of teaching and learning science.

In this chapter, we argue for developing a science teacher education, which embraces broad and inclusive approaches to teaching methodologies, in order to build a foundation that meets and considers the learner and his/her cultural background. This necessitates teacher educators to collaborate and learn from each other across different cultural settings, using different theoretical lenses and interpretive frames. By acknowledging and valuing the cultural background of the learner becoming teachers may impose more personalised and relevant learning and more frequent opportunities to discuss sustainability issues in a broader, culturally embedded perspective. On the global scales, the current devastating effects of climate change, biodiversity loss and the pandemic put education at large, in focus anew. In this perspective, science teacher education is one part, with a significant role to play in order to prepare becoming teachers to understand situated teaching practices and being able to adjust their teaching to learners from different cultural settings. This would mean not only to implement sustainability issues as topics in science education, but also to develop sustainable teaching strategies, a more sustainable science education.

## References

Anamuah-Mensah, J. (1998). Native science beliefs among Ghanaian students. *International Journal of Science Education*, *20*(1), 115–124.

Andersson, J., & Risberg, J. (2018). Embodying teaching: A body pedagogic study of a teacher's movement rhythm in the 'Sloyd' classroom. *Interchange*, *49*(2), 179–204.

Atwater, M. M. (1994). Research on cultural diversity in science classroom. In D. L. Gabel (Ed.), *Handbook on research on science teaching and learning*. New York, NY: Macmillan.

Breidlid, A. (2009). Culture, indigenous knowledge systems and sustainable development: A critical view of education in an African context. *International Journal of Educational Development*, *29*(2), 140–148.

Breytenbach, J. (2013). *A critical review of ICT-enabled development influencing the quality and quantity of South African tertiary education students*. Doctoral dissertation, University of Pretoria.

Chilisa, B. (2012). *Indigenous research methodologies*. Thousand Oaks, CA: SAGE.

Cobern, W. W. (1993). A co-operative Research Group for the study of culture and Scientific Education in developing countries. Paper presented at the UNESCO International conference on science education in developing countries, Jerusalem, Israel, 6 September 1992.

Creswell, J. W. (2009). *Research design: Qualitative and mixed methods approaches*. Thousand Oaks, CA: SAGE.

Englander, M. (2019). General knowledge claims in qualitative research. *The Humanistic Psychologist*, *47*(1), 1–14. doi: 10.1037/hum0000107

Engelhardt, P. V., Corpuz, E. G., Ozimek, D. J., & Rebello, N. S. (2003). Physics Education Conference. In *AIP Conference Proceedings* (Vol. 720, pp. 157–160).

Erickson, G. (2010). In the path of Linnaeus the development and nurturing of science educators for a complex world. In C. Linder, L. Östman, D. A. Roberts, P.-O. Wickman, G. Erickson, & A. MacKinnon (Eds.), *Exploring the landscape of scientific literacy* (pp. 207–222). Abingdon: Routledge.

Feinstein, N. W., & Waddington, D. I. (2020). Individual truth judgments or purposeful, collective sensemaking? Rethinking science education's response to the post-truth era. *Educational Psychologist*, *55*(3), 155–166.

Hager, P., & Johnsson, M. C. (2012). Collective learning practice. In *Practice, learning and change* (pp. 249–265). New York, NY: Springer.

Ingalls, L., Hammond, H., Dupoux, E., & Baeza, R. (2006). Teachers' cultural knowledge and understanding of American Indian students and their families: Impact of culture on a child's learning. *Rural Special Education Quarterly*, *25*(1), 16–24.

Inokoba, Preye & Adebowale, Adeyemi-Suenu & Perepreghabofa, Johnnie. (2010). The African metaphysical worldview and its prostrate condition of backwardness. *Journal of Human Ecology*. *29*(1), 23–31. doi: 10.1080/09709274.2010.11906245

Johnson-Glenberg, M. C., Birchfield, D., Tolentino, L., Koziupa. T. (2013). Collaborative embodied learning in mixed reality motion-capture environments: Two science studies. *Journal of Educational Psychology*, *106*(1), 86–104.

Leicht, A., Heiss, J., & Byun, W. J. (2018). *Issues and trends in education for sustainable development* (Vol. 5). Paris: UNESCO Publishing.

Merleau-Ponty, M. (1989). *Phenomenology of perception*. London: Routledge.

Munro, M. (2018). Principles for embodied learning approaches. *South African Theatre Journal*, *31*(1), 5–14.

Odhiambo, T. R. (1972). Understanding of science: The impact of the African view of the nature of science. In P. G. S. Gilbert, & M. N. Lovegrove (Eds.), *Science education in Africa*. London: Heinemann Educational Books Ltd.

Ododo, O. M. (2014). Influence of cultural practice-related misconceptions on achievement of senior secondary biology students in zone c of Benue State, Nigeria. *Journal of Education, Society and Behavioural Science*, 1703–1715.

Odora Hoppers, C. A. (2002). *Indigenous knowledge and the integration of knowledge systems: Towards a philosophy of articulation*. Claremont, South Africa: New Africa Books.

Ogunniyi, M. B. (1985). Problems of science education relative to the nature of scientific concepts and generalizations in developing countries. In F. M. A. Ukoh (Ed.), *What is science? Problems of teaching science and research in Nigerian universities*. Ibadan: Heinemann Education Books, Nigeria Ltd.

Ogunniyi, M. B. (1988). Adapting Western science to traditional African culture. *International Journal of Science Education*, *10*(1), 1–9.

Okebukola, P. (2002). *Beyond the stereotype to new trajectories in science teaching*. Kwali: Science Teachers Association of Nigeria.

Okere, M. I., & Keraro, F. N. (2002). Cultural influences on childhood science education. The Kenyan experience. *Egerton Journal, Humanities, Social Sciences and Education Series*, *4*, 141–153.

Okoye, B. E., & Okeke, O. C. (2007). Efficacy of eliminating superstitious beliefs strategy of achievement and knowledge retention in genetics among secondary school students. *Journal of the Science Teachers Association of Nigeria*, *42*(1), 73–77.

Ord, K., & Nuttall, J. (2016). Bodies of knowledge: The concept of embodiment as an alternative to theory/practice debates in the preparation of teachers. *Teaching and Teacher Education*, *60*, 355–362.

Ponelis, S. R. (2015). Using interpretive qualitative case studies for exploratory research in doctoral studies: A case of information systems research in small and medium enterprises. *International Journal of Doctoral Studies*, *10*(1), 535–550.

Säljö, R. (2009). Learning, theories of learning, and units of analysis in research. *Educational Psychologist*, *44*(3), 202–208.

Samba, R. M. O. (2003). Promoting students' understanding of photosynthesis and respiration: An evaluation of the potency of a conceptual change instructional strategy The Benue Valley. *Journal of Inter Disciplinary Studies*, *2*, 100–114.

Shizha, E. (2017). Indigenous knowledges and knowledge codification in the knowledge economy. In *Handbook of research on theoretical perspectives on indigenous knowledge systems in developing countries* (pp. 267–288). Hershey, PA: IGI Global.

Shumba, O. (1995). Interaction of traditional socio-cultural variables with instructional ideology and knowledge of the nature of science: A review of the literature with implications for Zimbabwe. *Zimbabwe Journal of Educational Research*, *7*(1), 23–56.

Simayi, A., & Webb, P. (2019). *Asibizi: Teaching human reproduction in rural Eastern Cape schools* (No. 2284). Manchester: EasyChair.

Smyrnaiou, Z., Sotiriou, M., Georgakopoulou, E., & Papadopoulou, O. (2016). Connecting Embodied Learning in educational practice to the realisation of science educational scenarios through performing arts. In *International Conference "Inspiring Science Education"*, Athens, 22–24 April 2016, pp. 37–45.

Snively, G., & Corsiglia, J. (2001). Discovering indigenous science: Implications for science education. *Science Education*, *85*, 6–34. https://doi-org.ezproxy.ub.gu.se/10.1002/1098-237X(200101)85:1≤6::AID-SCE3≥3.0.CO;2-R

Stake, R. E. (2013). *Multiple case study analysis.* New York, NY: Guilford Press.

Stolz, S. A. (2015). Embodied learning. *Educational Philosophy and Theory*, *47*(5), 474–487.

Sund, L., Quennerstedt, M., & Öhman, M. (2019). The embodied social studies classroom–Repositioning the body in the social sciences in school. *Cogent Education*, *6*(1), 1569350.

Sullivan, J. V. (2018). Learning and embodied cognition: A review and proposal. *Psychology Learning & Teaching*, *17*(2), 128–143.

Swedish National Agency of Education. (2017) *Kommentarmaterial till kursplanen i fysik [Supplement to the syllabus in Physics]*. Retrieved from https://www.skolverket.se/publikationer?id=3789

Thorburn, M. (2008). Articulating a Merleau-Pontian phenomenology of physical education: The quest for active student engagement and authentic assessment in high-stakes examination awards. *European Physical Education Review*, *14*(2), 263–280.

UNESCO. (2021). *UN Decade of ESD.* United Nations Educational, Scientific and Cultural Organiation (UNESCO). https://en.unesco.org/themes/education-sustainable-development/what-is-esd/un-decade-of-esd

UNESCO. (2017). Education for Sustainable Development Goals. Learning objectives. Paris: UNESCO. Retrieved from https://unesdoc.unesco.org/ark:/48223/pf0000247444

Wals, A. (2009). *Learning for a sustainable world: Review of contexts and structures for education for sustainable development 2009.* Paris. UNESCO.

Yin, R. K. (2015). *Qualitative research from start to finish*. New York, NY: Guilford Publications.

Yip, D. Y. (2001). Promoting the development of a conceptual change model of science instruction in prospective secondary biology teachers. *International Journal of Science Education*, *23*(7), 755–770.

Zidny, R., Sjöström, J., & Eilks, I. (2020). A multi-perspective reflection on how indigenous knowledge and related ideas can improve science education for sustainability. *Science & Education*, *29*(1), 145–185. doi: https://doi.org/10.1007/s11191-019-00100-x

# 9 Digital competence with respect to ESD for science and technology student teachers

## A study of a South African and a Swedish university

*Anne-Marie Cederqvist, Ann-Marie von Otter, Portia Kavai and Samuel Khoza*

### Introduction

The Fourth Industrial Revolution (4IR), which is characterised by digital technologies such as artificial intelligence, genetic engineering and 3D printing of organs, are fusing the physical, digital and biological worlds and are changing the way we see humans and society (Schwab, 2017). The potential of the 4IR is followed by great expectations of how digital technology will be the contributing force that will make it possible to achieve the sustainable development goals (SDGs). Hence, there is an increased emphasis in education for sustainable development (ESD) on digital competence and skills related to digital technology. However, from a global perspective, there are differences between countries across the world such as access to digital technology as well as access to digital public infrastructure, which may affect to what extent citizens can take part in the potential of the 4IR. Despite these differences, teacher education across the world is facing similar changes that need to be taken into account when preparing student teachers, i.e., changes regarding the ambition in compulsory school curricula of equipping learners with digital competence. In cases of South Africa and Sweden, as well as in many other countries, digital knowledge and skills have become important elements in science and technology education to meet the demands of the digital society and the demands for trained workforce that can compete in the globalised digitised environment. Hence, science and technology teacher education programmes in both South Africa and Sweden are facing changes in preparing future teachers to provide learners with digital competence. However, when teaching digital competence, sustainability-related content needs to be incorporated together with teaching and learning approaches that stimulate and educate individuals as agents of change. That is, science and technology teacher education need to prepare future teachers to educate learners to take action for sustainable development

DOI: 10.4324/9781003190158-12

with the help of digital technologies of the 4IR. However, issues with respect to ESD and digital competence are rarely discussed in relation to each other. Therefore, this chapter presents the collaboration and exchange of experiences of teacher educators (TEs) in South Africa and Sweden, on the relationship between ESD and digital competence, as well as discuss how to form a teacher education that meets the requirements of science and technology teachers in their future professions. South Africa and Sweden are examples of countries that are both similar and different at the same time in as far as their teacher training programmes are coined. On the one hand, there are differences for instance in access to digital technologies as well as access to digital public infrastructure. On the other hand, both countries have recently implemented several aspects of digital competence in the curricula for compulsory school education. Accordingly, the point of departure in this study was the syllabuses in science and technology education for compulsory schools in South Africa and Sweden. The curricula of the two countries (or syllabuses as it is known in Sweden) were analysed and compared to establish possible similarities and differences that could inform this study on the extent the teacher education institutions of the two countries prepare science and technology student teachers for teaching digital competence with respect to ESD and the possible gaps. The research questions that guide the study are:

- In what ways are ESD incorporated in the science and technology syllabuses covering digital competence in South Africa and Sweden?
- What are the similarities and differences between the South African and the Swedish science and technology syllabuses in covering digital competence with respect to ESD?
- To what extent do the science and technology teacher education in the two institutions cover the content of digital competence in relation to sustainable development and what more should be done?

## Background

Of late, there has been a buzzword of "4th Industrial Revolution (4IR)". There is an acclaimed potential of the 4IR to improve the quality of life for all people in the world. Many countries are using new products and services, developed to take advantage of digital technology. Workplaces and organisations are becoming "smarter" and more efficient, and there is an increased use of digital technology to improve people's lives in terms of economic growth, education, health care as well as equality aspects. Furthermore, the potential of the 4IR may help people better prepare for natural disasters and reduce climate threats.

The 4IR is characterised by advances in Artificial Intelligence (AI), robotics, the Internet of Things (IoT), 3D printing, genetic engineering, quantum computing and other technologies that are merging together digital, biological and physical worlds (Penprase, 2018). Digital technology is the common force behind many products and services that quickly have become essential in everyday life, such as GPS systems that suggest the fastest route to a destination, autonomous vehicles, voice-activated virtual assistants like Apple's Siri, individual Netflix recommendations and Facebook's recognition abilities.

In the book, The Fourth Industrial Revolution, Klaus Schwab provides a primer on the 4IR and what it will bring: "The changes are so profound that, from the perspective of human history, there has never been a time of greater promise or potential peril" (Schwab, 2017, p. 2). The digital transformation of our society has had a huge impact on the labour market, economy, education and our social lives; and digital technology has become a central part in our everyday lives. Almost everything we take part in requires digital skills and knowledge to navigate through the day. However, individuals do not only need skills and knowledge about how to use digital technology and what possibilities it may create, they also need abilities to critically reflect on issues concerning sustainability and how digital technology affects humanity, societies and the environment. It is therefore undeniable that education plays an important role to develop individuals' digital competence and help them become more responsible and skilled critical consumers, as well as producers. Hence, it is necessary for education to reflect on pedagogies and strategies for providing digital competence that encompasses sustainability issues to meet both the promises and the challenges of the 4IR. The crucial question therefore that this chapter asks is, are the teacher education institutions of the two countries adequately preparing the science and technology student teachers for digital competence in the era of Anthropocene?

## ESD and digital competence

There has been an increased emphasis in ESD on skills to meet the demands of the 21st century such as managing Information and Communication Technologies (ICT) (Rieckmann, 2018). Voices are raised about how digital technology, because of its critical trans-sectoral nature, will be the contributing force that will make it possible to achieve the SDGs. Digital technology and new innovative technological solutions will have a key role when it comes to developing economic growth in businesses and societies, promoting better processes and production patterns in industry, promoting better consumption patterns and resource optimisation in water and energy consumption, and be important in the work of environmental protection and to prevent climate changes (Global e-Sustainability Initiative, 2017). Digital technology will also give the possibility to increase equality in several areas since more people will get access to many basic services provided by Internet and mobile technology such as education, health care and public information and services in communities (United Nations, 2019). However, there should be awareness that digital technology comes with consequences such as e-waste and an increased use of energy, and digital technology will affect daily life and the labour market. While digital technology will result in job losses in many sectors, it will at the same time create new jobs that do not exist today (Berkout & Hertin, 2004; United Nations, 2019). To meet these opportunities and challenges with digital technology, skills and knowledge related to what is defined as digital competence among citisens will be required. Accordingly, the provision of high-quality education will be an important key (Atchoarena et al., 2017).

ESD is defined as education, which provides opportunities for learners to acquire knowledge, skills, attitudes and values that are necessary for shaping a sustainable future (Rieckmann, 2018). As the advocacy for the 4IR increases, so should the advocacy for ESD to ensure that the same digital users are not jeopardising sustainability but are responsible citizens. Thus, digital technology alone will not be enough to achieve the SDGs; it comes with a closely intertwined mindset, which needs to be addressed in teaching. This mindset is related to what Leicht, Heiss and Byun (2018) describe as the socio-emotional and behavioural learning objectives, which will provide learners with values, attitudes and action competencies on how digital technology is going to be developed and used in an innovative way, and what consequences it may have on humans, society and the environment. A central issue is then how this mindset is going to be a common way of thinking and integrated into teaching when it comes to the development and use of digital technology in the future.

If sustainable development principles are integrated into education, from preschool to tertiary education, then education itself could be a powerful tool to achieve all of the SDGs (Rieckmann, 2018). ESD should then equip learners with both the knowledge about the SDGs and with the skills and competence to achieve the SDGs. However, to be able to realise this, education in itself needs to be transformed and employ new approaches where human well-being and global development are in focus and to educate individuals as agents of change (Leicht et al., 2018).

Digital competence (DC) generally refers to skills and knowledge that are needed to be able to learn and act in the digitalised society (Ilomäki, Paavola, Lakkala, & Kantosalo, 2016). By taking several frameworks into account Ferrari, Punie and Redecker (2012) define digital competence as follows:

> Digital Competence is the set of knowledge, skills, attitudes, abilities, strategies and awareness that is required when using ICT and digital media to perform tasks; solve problems; communicate; manage information; behave in an ethical and responsible way; collaborate; create and share content and knowledge for work, leisure, participation, learning, socialising, empowerment and consumerism
>
> (Ferrari et al., 2012, p. 84)

The above definition emphasises that digital competence is a need that every individual should consider having in order to cope with the change that the world is becoming. The concept digital competence is multifaceted and may have different meanings depending on the context where it is used. This implies that even if digital competence has a general meaning for what is required for being able to navigate in a digital society, it may have a specific meaning in an educational context that is necessary to discuss. Instefjord and Munthe (2016) argue that digital competence for teachers is actually a more complex set of skills and knowledge, compared to other areas in society. There is a twofold dimension of teachers' digital competence since teachers are not only supposed to gain digital competence for their own use of digital technology, they are also going to educate and instruct learners regarding skills and knowledge related to digital competence. This implies that a pedagogical compatibility to the different knowledge areas of digital competence is required (Instefjord & Munthe, 2016). Krumsvik (2011) defines teachers' and TE's digital competence as follows:

> Digital competence is the teacher/TEs' proficiency in using ICT in a professional context with good pedagogic-didactic judgement and his or her awareness of its implications for learning strategies and the digital Bildung of pupils and students
>
> (Krumsvik, 2011, pp. 44–45)

The pedagogic didactic-judgements are a part of a teachers' daily work and is based on the teachers' pedagogical content knowledge (PCK) regarding the subject to teach (Shulman, 1987). Therefore, it is important in teacher education to provide a pedagogical framework, which includes a didactical content, which meets the requirements of the digital competence that teachers need (Krumsvik, 2011).

### *The DigComp 2.1: The Digital Competence Framework for Citizens*

DigComp 2.1: The Digital Competence Framework for Citizens (Carretero, Vuorikari, & Punie, 2017), **which is summarised from its competence areas** in Table 9.1, has become a reference for developing initiatives for implementing digital competence in the learning and employment field, mainly in member states of the European Union. However, it may also be used inversely as a framework for investigating policy documents. DigComp 2.1 framework can be seen as a tool for developing citizen's digital competence. Hence, DigiComp 2.1 guides education, as well as training practices to improve workplace performance, on how to provide citizens with digital competencies and skills required in today's society for personal development, social inclusion and employment.

The DigComp 2.1 framework is then what guides the European Framework for the Digital Competence of Educators: DigCompEdu (Redecker, 2017) shown in Figure 9.1. The DigCompEdu is describing digital competence in relation to education specifically and works as a reference for educators to support development of digital competence in educational contexts at all levels of education (i.e., from early childhood to higher education). The framework comprises digital competence with respect to digital technologies used for enhancing education as well as for developing learners' digital competence. In this study, the

*Table 9.1* The Competence Areas of DigComp 2.1: The Digital Competence Framework for Citizens

| *Information and data literacy* | *Communication and collaboration* | *Digital content creation* | *Safety* | *Problem solving* |
|---|---|---|---|---|
| Browsing, searching, filtering data, information and digital content<br>Evaluating data, information and digital content<br>Managing data, information and digital content | Interacting through digital technologies<br>Sharing through digital technologies<br>Engaging in citizenship through digital technologies<br>Collaborating through digital technologies<br>Netiquette<br>Managing digital identity | Developing digital content<br>Integrating and re-elaborating digital content<br>Copyright and licenses<br>Programming | Protecting devices<br>Protecting personal data and privacy<br>Protecting health and well-being<br>Protecting the environment | Solving technical problems<br>Identifying needs and technological responses<br>Creatively using digital technologies<br>Identifying digital competence gaps |

Source: Adapted from Carretero et al. (2017).

research interest is directed towards the implementation of digital competence in compulsory school, and the preparation of student teachers for taking part in this development. Therefore, this study was guided by the five digital competencies that are specified in DigComp 2.1 (Table 9.1), which also corresponds to area 6 in the DigCompEdu framework shown in Figure 9.1.

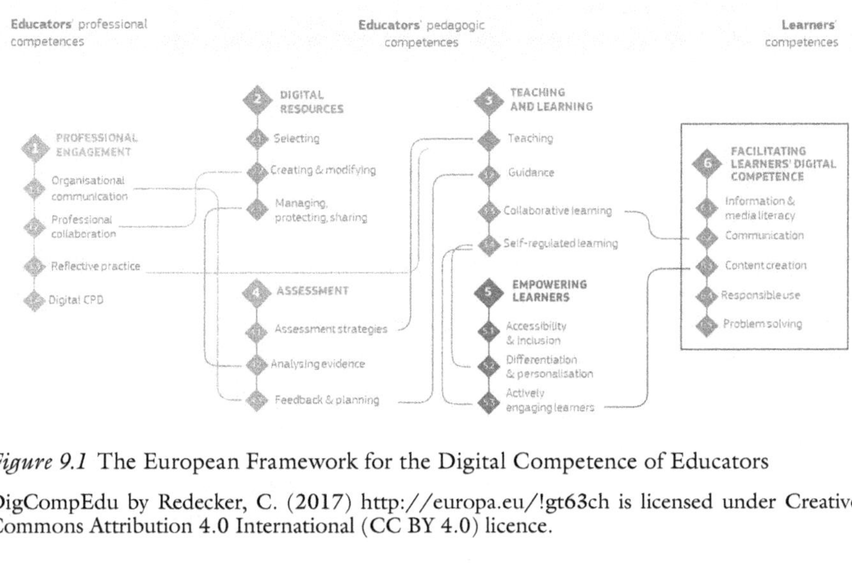


*Figure 9.1* The European Framework for the Digital Competence of Educators

1 *Information and media literacy*

To incorporate learning activities, assignments and assessments which require learners to articulate information needs, to find information and resources in digital environments, to organise, process, analyse and interpret information, and to compare and critically evaluate the credibility and reliability of information and its sources.

2 *Communication and collaboration*

To incorporate learning activities, assignments and assessments which require learners to use digital technologies effectively and responsibly for communication, collaboration and civic participation.

3 *Digital content creation*

To incorporate learning activities, assignments and assessments which require learners to express themselves through digital means and to modify and create digital content in different formats. To teach learners how copyright and licenses apply to digital content, how to reference sources and attribute licenses.

4 *Safety*

To take measures to ensure learners' physical, psychological and social well-being while using digital technologies. This would be to empower learners to manage risks and use digital technologies safely and responsibly. The teachers may integrate ESD on this aspect instilling ethics and responsibility of the learners regarding e-waste and over-use of energy

5 *Problem solving*

To incorporate learning activities, assignments and assessments which require learners to identify and solve technical problems, or to transfer technological knowledge creatively to new situations.

The sixth concept/area shown in Figure 9.1 complimented the above digital competencies. This was deemed relevant because area 6 in the figure aims to facilitate learners' digital competence.

The presented digital competencies go beyond the bare use of digital technology for facilitating teaching and learning. They also include how digital technology can be used in a creative and responsible way to solve problems as well as provide possibilities for learners to participate and act, by the benefit from digital technologies. These competencies should also be seen as intertwined with ESD and the idea of providing digital competence that makes it possible for taking action for sustainable development with the help of digital technologies. In this study, the DigComp 2.1 framework is used on basis of the presented competencies, which are seen as universal. That is, the competencies are not seen as unique and specifically related to citizens within EU, but to all citizens in the world. Thus, the DigComp 2.1 framework guides this study when analysing the curricula in compulsory schools, in both South Africa and Sweden. The aim of the study was to investigate and discuss the relationship between ESD and digital competence in science and technology teacher education in the two countries. Our point of departure was investigating the curricula for compulsory school in South Africa and Sweden since these school curricula are the ones that inform the teacher education curricula. Further, we discussed the extent to which the teacher education institutions of the two countries prepare science and technology student teachers for teaching digital competence with respect to ESD and the possible gaps that exist in the systems. A brief description of the context concerning policies and documents that guide the development of teacher education courses in South Africa and Sweden is presented.

## Teacher education in the Swedish context

Teacher education in Sweden takes place at the universities that have teacher education departments. Since autumn 2011, there are four different vocational degrees: preschool teachers which carries 210 credits, elementary teachers with 240 credits, subject teachers with credits that are between 270 and 330 credits. The four vocational qualifications of teacher education are regulated in the Higher Education Law. To be able to apply for a teacher-training course, a basic qualification at the basic level can be fulfilled in various ways, but the most common one is a completed education at the upper secondary school level. It is the Swedish Higher Education Law that determines the main focus on teacher education. However, it is the head of the university who sets out what the syllabus should look like, and then the respective teacher education institution sets the content and prerequisites based on a link collection that shows paths to national and local governance documents as well as to university common resources and contacts.

The University of Gothenburg Sweden is actively working to include and collaborate with students on all sustainability and digitisation issues. The University of Gothenburg strives to increase the integration of sustainable development in education and applies sustainability labelling of courses and programmes. The marking is based on the learning objectives of the syllabus and is assessed based on university-wide criteria that have been established by the Board of Education. Digital competence in the perspective of teacher education in general also means the use of digital technology in various ways to improve teaching. However, a previous study shows (see Sofkova Hashemi et al., 2019) that digital technology and how to use it in teaching are not so well integrated into teacher education in Sweden.

### *Digital competence in the Swedish compulsory curriculum*

Since the compulsory school (primary and lower secondary) curricula inform the teacher education curricula, it is important to consider it to enable a better understanding of the teacher education curricula. In Sweden, a revised curriculum with an emphasis on digital competence was implemented in compulsory schools in 2018 (Skolverket, 2018). The revised curriculum is based on how the Digitization Commission [Digitaliseringskommissionen] (2014) describes digital competence: "Digital competence consists in the extent to which one is familiar with digital tools and services and is able to monitor digital development and its impact on one's life". Four aspects of digital competence are highlighted in the revised curriculum that will affect teaching in all subjects: to understand how digital technology affect the society; to be able to use and understand digital tools and media; to have a critical and responsible approach about possibilities and risks in relation to digital technology; and being able to solve problems and put ideas into action with digital technology (Skolverket, 2017). Programming has been implemented as part of digital competence where programming should be viewed in a wider perspective and thus included in all aspects of digital competence such as writing code, solving problems, and in relation to democratic dimensions.

The changes in the Swedish curriculum regarding digital competence were motivated by arguments that the content of teaching should be adapted to the knowledge the learners need to be able to act as active and reflective citizens in the increasingly digitised society. Another argument was that there are major differences between schools regarding the use of digital technology in teaching and that there is a need for equality in the school system. The four aspects of digital competence above can all be found throughout the Swedish curriculum for the compulsory school, both in general parts and in the syllabuses for specific subjects (see Skolverket, 2018). Since the aim of this study is to compare the South African and Swedish teacher education regarding digital competence in the subjects of science and technology and with respect to ESD, it is deemed necessary to explore what the syllabuses for science and technology in the two countries say with respect to digital competence and ESD.

### *Digital competence in the Swedish syllabuses for science and technology*

Teaching in science education is expected to develop learners' ability to reflect on what opportunities and constraints that the use of digital technologies may bring, and in what ways these may affect the environment. Furthermore, the syllabus for science education describes that: "As part of science systematic studies, learners should be given opportunities to develop skills in the use of both digital tools and other equipment" (Skolverket, 2018, p. 166). Skolverket (2017) suggests that the digital tools should be used in teaching to clarify phenomena, to document science studies and to provide opportunities for learners to develop skills for using the digital tools. The digital tools are suggested to be used to vary teaching and work formatively. In addition, teaching in science education also includes a source-critical review of information and arguments in digital media (Skolverket, 2018).

In technology education, the overarching goal is to develop learners' understanding of the human constructed world and thus also of the digital world (Skolverket, 2018). To gain this understanding, learners need to learn about the phenomenon of digital technology, but they also need to learn how to use digital tools within the digital world (Skolverket, 2017). In addition, the teaching of technology should develop learners' understanding of how technology affects humans, the society and the environment. According to Skolverket, the revised syllabus in technology highlights several areas that can be related to the digital world and digital competence. The main areas are: what computers are used for and some of the basic parts of a computer; technological solutions in communication and information technologies such as computers; the Internet; technological solutions that use electronics and how they can be programmed; how digital tools can provide support in technological development work by producing drawings and simulations; and documentation in the form of digital sketches, drawings and digital models (see Skolverket, 2018, pp. 297–299). These curriculum changes need to be seen in the light of the overarching goal where "the school is responsible for ensuring that each learner on completing compulsory school has obtained knowledge about the prerequisites for a good environment and sustainable development" (Skolverket, 2018, p. 12).

## Teacher education in the South African context

Teacher training in South Africa happens mostly in the Bachelor of Education (BEd) degree. This is a four-year degree with 480 credits. However, according to the Minimum Requirements for Teacher Education Qualifications (MRTEQ) (Department of Higher Education and Training (DHET), 2015), the BEd degree is divided into four different groups according to the age of the learners to be taught which are generally referred to as learners in the South African context, namely, Foundation phase, Intermediate phase, Senior phase (upper primary and lower secondary) and Further Education and Training (FET) phase (upper secondary). The BEd programme aims at providing a well-rounded education that equips graduates with the required subject content knowledge base, educational theory and methodology that would enable them to demonstrate competence and responsibility as academically and professionally qualified beginner teachers (Department of Higher Education and Training (DHET), 2015).

MRTEQ implicitly alludes to the fact that the world is busy with digital competence conversation and as such, tertiary institutions may infuse digital competence in their teaching to prepare for globally competitive future teachers. However, since the teacher qualifications in South Africa are aligned to MRTEQ, any other addition that is not explicit in this document is most likely to be disregarded, creating a gap in ICT integration in teacher education.

### *Digital competence in the South African compulsory curriculum*

The South African curriculum has been changing since the dawn of democracy. During the late 1990s, the post-apartheid government introduced the Outcomes Based Education as part of the Curriculum 2005 project (see Taylor & Vinjevold, 1999), by adapting the Australian education system (Donnelly, 2007). Since then the curriculum was changed in 2008, and later amended in

2013 to the current Curriculum and Assessment Policy Statement (CAPS) document (du Plessis & Marais, 2015). This was an endeavour to produce quality learners who would take part in global activities since the world is moving with the times. The question that is important to ask is the level of preparedness of the implementers of the ever-changing curricula, the teachers. As a country, South Africa needs a trained workforce that can compete in the globalised digitised environment.

The main aim of the National Curriculum Statement (NCS) and other specific aims in science education implicitly allude to ESD per topic leaving it open to teachers on how far they would go according to the grade taught. In 2015 the Education Member of Executive Council (MEC) for Gauteng province came up with an innovative idea of what he termed "paperless schools" in a bid to enforce the use of technology in the teaching of all subjects in response to emerging technologies and digitisation era. Smartboards, visualisers, projectors, teacher laptops, learner tablets and free access Wi-Fi were provided, initially to the formerly disadvantaged schools which have learners of a low socio-economic status. Currently vision 2020 aims to expand this digitisation beyond just one province. This is to be fulfilled by the introduction of a coding and robotics curricula to be included from Grades R-9 in 2021. According to the minister "The curricula will provide learners with an understanding of coding and robotics and will develop their skills and competencies to prepare them for the 4IR". The vision for this curriculum is to ensure that the schooling system produces learners with the foundations for future work and equip them with skills for a changing world, irrespective of their socio-economic background. The coding curriculum will develop learners' ability to: solve problems, think critically and work collaboratively and creatively; function in a digital and information-driven world; apply digital and ICT skills; to transfer these skills to solve everyday problems. The robotics curriculum will have a strong foundation in engineering and will enable learners to build and operate robots through programming code. Several new technology subjects and specialisations would be introduced with subjects like Technical Mathematics, Technical sciences already being done in some schools since 2016, other subjects will include Maritime sciences, Aviation studies, Mining sciences and Aquaponics and in each one of them, there is a requirement for digitisation and inference on ESD.

### *Digital competence in the South African syllabuses for science and technology*

In the South African context, technology is two-pronged, one of which is technology as a subject that is associated with engineering taught in Grades 7–9 and the second one is the technology linked to ICT and/or digital competence both of which are focused on, in this chapter. The South African CAPS document (Department of Basic Education, 2011) emphasises among other aims the ability for learners to communicate effectively using visual, symbolic and/or language skills in various modes; use science and technology effectively and critically showing responsibility towards the environment and the health of others, and demonstrate an understanding of the world as a set of related systems by recognising that problem-solving contexts do not exist in isolation. This main aim and other specific aims in the science education implicitly allude to ESD and use of technology in science education per topic leaving it open to teachers on how far they would go according to the grade taught.

Another curriculum, which would integrate ICT, has been drafted and is to be introduced to the schools. In this new curriculum, ICT involves the use of computers, applications and Internet technologies to enable users to communicate, create, store, distribute and manage information. The curriculum is offered as an employability skill by the school that may decide on the format of delivery. Schools may use fixed periods or may deal with the topics for ICT by integrating relevant ICT topics across the curriculum in General and or Technical and Vocational Education. It is compulsory to offer the skills, but the delivery method may differ to suit the school context.

As of 2018–2019 the South African Department of Education in collaboration with other stakeholders like Microsoft, Vodacom and Standard bank have been working together in developing a new curriculum focusing explicitly on digital competence. The South African National Department of Education's intention was to pilot Coding and Robotics in Grades R-3, in 200 selected schools across the country in 2020 – with a plan to implement it fully by 2021/2022 across the grades from Grade R to Grade 9. This is one of the steps taken by the government to improve education in the country and to equip learners with the required skills for the 4IR. The drafting of the Coding and Robotics Curriculum for Grade R to Grade 9 concluded in 2020. The Draft Curriculum has been subjected to approval processes within the sector, after which the Minister of education have gazetted the curriculum for public comment. All these changes are evidence that the education system leaders acknowledge the career and livelihood opportunities presented by the 4IR but the education system still needs to be transformed in practice even more and countrywide if the country is to meet them and remain relevant in the digital world.

Based on the new directives in the curricula for compulsory school, in both South Africa and Sweden, it is necessary for teacher education to reflect on pedagogies and strategies for providing teacher education to meet the requirements of digital competence, with respect to ESD. This encompasses knowledge, skills, attitudes and values that are necessary for shaping a sustainable future to meet both the promises and the challenges of the 4IR.

## Methodology

This qualitative case study had data collected through document analysis of the Swedish compulsory school curricula and the South African National Curriculum statement. The study includes content analysis of the syllabuses within the subjects, science and technology for compulsory school in South Africa and Sweden.

For the Swedish context, the revised curriculum from 2018 was used. For the SA context, the MRTEQ document was used as a guideline because all the teacher education qualifications in SA are aligned to it and recognised in the process. The spotlight was therefore on the SA BEd degree.

The purpose of analysing these documents was to map how digital competence and ESD are presented in the policy documents, within the framework of compulsory school in South Africa and Sweden. The Swedish school offers a 9-year compulsory school and the South African school offers 9-year compulsory school, which include foundation, intermediate and senior phase. The two subject areas that were focused on were science and technology for learners in compulsory school Grades 1–9 aged 7–16. Analysing the documents gave enlightenment on the digital content and ESD that student teachers should encounter in the two countries' teacher education programmes with regard to science and technology education in compulsory school. By comparing the syllabuses for the two countries, we were able to find similarities and differences that could broaden and enrich the discussion on how to form a teacher education that meets the requirements of science and technology teachers in their future profession, in both South Africa and Sweden.

The data was analysed using the DigComp 2.1 framework based on the following categories: data and information knowledge, communication and collaboration, digital content creation, security and problem solving. In the analysis, each syllabus of science and technology was analysed to identify content that emphasises the categories of the DigComp 2.1 framework for digital competence. Since the revised South Africa curriculum has not yet been implemented, the analysis involves both the current science and technology syllabi and the drafted curriculum based on the DBE Concept Document: Introducing Digital Skills for all into GET (Labuscagne, 2018).

## Findings and discussion

The question is: how prepared are science and technology teacher education programmes in South Africa and Sweden in the integration of digital technology and equipping the student teachers with digital competence and ESD integration in the content teaching? To answer this question, the point of departure is the analysis of syllabuses in science and technology education for compulsory school in South Africa and Sweden. The purpose of the analysis is to understand in what ways the syllabuses address content with respect to digital competence based on the DigComp 2.1 framework and its categories: information and data literacy, communication and collaboration, digital content creation, safety and problem solving, as an analytic tool. The findings from the syllabuses' analysis are then discussed in relation to the extent digital competence and ESD are currently covered in the two teacher education institutions. Further, the findings are discussed on what ways teacher education programmes may need to be transformed to meet the realities the teachers will face in their future profession.

### *The ways digital competence is covered in the syllabuses for science and technology*

The findings are based on the analysis of the Swedish curriculum for compulsory school and the current South African curriculum regarding science and technology education, together with the proposed changes in the ongoing project of revising the South African curricula with respect to digital competence. The content analysis of the mentioned syllabuses for science and technology education in each country's curricula shows that the areas of digital competencies regarding the categories in the DigComp 2.1 (Carretero et al., 2017) are covered to some extent. In the syllabuses for technology, both in South Africa and Sweden, there is a broader coverage on digital competencies probably since the subject itself is more naturally linked to digital technology.

### *The syllabuses for technology*

With respect to the category *information and data literacy,* the content of the Swedish syllabus is focusing on technicalities with respect to the use of digital technology for documentation and writing reports, and how different technological solutions or technological systems work for exchanging information. Critical skills concerning browsing or searching for data, information and other digital content seem to be excluded. In the South African syllabus, the category *information and data literacy*, in the current version, is covered by content directed towards competencies and skills for managing data with respect to digital content creation. In the revised syllabus, this has been extended to also include the awareness of using digital technology in an appropriate way to access, integrate and evaluate information. Thus, the content in the syllabuses of both countries seems to be covering the management and use of digital technologies and digital content such as tools to access information and writing reports. However, there are no explicit formulations concerning the critical awareness that are necessary when managing digital content.

The content *communication and collaboration* is in the Swedish syllabus directed towards how digital technology changes the conditions for different occupations, and also the entire society, which may be related to interacting, sharing and collaborating through digital technology as well as engaging in citizenship through digital technologies. However, the content points more towards that learners should be aware of how digital technology affects society, and not towards learners' own engagement in interacting, sharing and collaborating. Further, the sub-categories netiquette and managing digital identity are not highlighted. The current South African syllabus points towards the content of Internet and e-communication. The revised syllabus expands this content by focusing on how to use the Internet and other ICT-tools for achieving particular purposes and to solve real-life problems, which may be related to interacting, sharing and collaborating through digital technology as well as engaging in citizenship through digital technologies. In addition, ethical and legal issues related to the access and the use of the Internet are highlighted where specifically bullying and trolling are mentioned, which is also pointing towards issues concerning *safety* and the protection of health and well-being. The South African syllabuses are more explicit regarding the use of Internet and ICT-tools for specific purposes that also include netiquette and individual well-being. On the other hand, the Swedish syllabus is more explicit regarding the way digital technologies change the conditions for human and society, which alludes to ESD.

*Digital content creation* including developing, integrating and creating digital technology where programming is a part seems to be well covered in the Swedish syllabus where also mankind's opportunities to create technological solutions that contribute to sustainable development are highlighted. The *digital content creation* category is closely related to the *problem-solving* category. Both these categories encompass knowledge about common technological systems at home and society and knowledge about technological solutions where mechanical and digital technology work together including the designing and programming of technological solutions. In this context, issues concerning consequences of technological choices and their advantages and disadvantages are mentioned as important parts of the design process. In addition, these categories also cover the structural and functional nature of computers and the use of digital tools to produce, e.g., drawings as a part of technological development work. Also, the South African syllabuses covers various content of *digital content creation*. The current syllabus includes producing spreadsheets and presentations, web design and application development. In addition, the revised syllabus holds content such as basic bitmap graphics and vector editing and highlights knowledge on how to use digital tools for solving problems. Thus, this content is also closely intertwined with *problem solving* with an explicit focus on computational thinking skills including problem-solving techniques, algorithms, coding and robotics. Furthermore, the revised content also includes troubleshooting of hardware and software. However, the content does not cover issues concerning the identification of needs and technological responses, which are important parts in the problem-solving process. Both the South African and the Swedish syllabuses cover digital competencies related to programming and solving problems. In addition, the Swedish syllabus explicitly mentions the opportunities of creating technological solutions that will contribute to sustainable development and brings up reflection on issues concerning consequences of technology.

The focus on *safety* is covered to a great extent in the Swedish syllabus of technology. Transferring information in digital environments, using the Internet and storing and protecting data are highlighted. Also issues concerning protection of health and well-being, and the environment are highlighted in terms of risks and limitations of the Internet and other global technological systems, together with issues concerning consequences of choices from ecological, ethical and social perspectives, both from a consumer and a producer perspective. When it comes to *safety* in the South African syllabuses, ethical and legal issues are highlighted with respect to access and use of the Internet, specifically bullying and trolling are directed towards individuals' health and well-being. However, neither the current syllabus nor the revised syllabus explicitly mentions anything about protecting the environment.

### *The syllabuses for science education*

In the syllabuses for science education we can see that the content concerning digital competence is mainly addressing *information and data literacy* in terms of integration and critical examination of information, arguments and sources found in digital media such as articles in newspapers and films and other digital sources. Also, content concerning the category of digital *content creation* is found; learners are supposed to document studies by using diagrams, pictures and written reports produced with digital tools. In addition, digital tools are also used in science education in a problem-solving manner as a support in systematic studies and field studies for modelling, planning, execution and evaluation where the use of digital instruments for measuring and registering properties in the environment is included.

### *The ways ESD is incorporated in the syllabuses covering digital competence*

With respect to the South African and the Swedish syllabuses including the new digital competence curricula which is still being piloted, findings show that the syllabuses for technology are covering all the categories of digital competence to a great extent and more clearly. In science education, the coverage is lower, the focus is mainly on integration and critical examination of information, arguments and sources found in digital media, and for documentation, measuring and registering properties in the environment by using digital tools. There are a few explicit formulations that are directed towards sustainability issues concerning digital technology, mainly in the technology syllabuses concerning digital content creation and safety issues. The highlighting of programming and solving problems and how this will provide opportunities for creating technological solutions that will contribute to sustainable development are in line with the arguments of how high-quality education including digital competencies are important for achieving the SDGs (Atchoarena et al., 2017; Rieckmann, 2018). It is of importance that these issues are made explicit also in teacher education for answering the didactical question of why learning about digital technologies is important. Furthermore, issues concerning consequences of technology, and what advantages and disadvantages they may have, are important sustainability issues that are highlighted in the technology syllabuses. The protection of health and well-being and the environment are important to incorporate when it comes to digital technology. The findings show that safety issues are brought up with respect to the use of the Internet and in what way this affects individuals' health and well-being over time. Furthermore, issues concerning risks and limitations of the Internet and other global technological systems, together with issues concerning consequences of choices from ecological, ethical and social perspectives are incorporated in the syllabuses for technology. These issues are also suggested by Schwab (2017) as necessary to reflect upon in the time of 4IR. However, Berkout and Hertin (2004) suggest that perceptions of environmental issues have become more dependent on digital technologies and mediated through digital instruments. Hence, there is required knowledge of how to manage these instruments as well as to interpret their results.

Leicht et al. (2018) argue for the importance of developing learners' socio-emotional and behavioural skills including values, attitudes and action competencies where critical skills necessary for managing and creating the digital technology may be included. As Leicht et al. state, the digital technology in itself will not be sufficient to achieve the SDGs, the digital technology need to be developed and used in an innovative way, where it is necessary to critical reflect on consequences of digital technology such as e-waste and an increased use of energy, and how digital technology will affect daily life and the labour market (Berkout & Hertin, 2004; United Nations, 2019). Although, issues concerning critical skills concerning digital content are explicitly described in the science syllabuses, these are more directed towards critical reflection on digital information rather than on the technology itself. In the technology syllabuses, there are some implicit formulations that may direct towards critical reflection on digital technologies. However, if teachers are going to be agents of change as Leicht et al. suggest, these issues need to be highlighted explicitly, preferably in teacher education. Thus, teacher education needs to include knowledge about the SDGs when delivering the digital competence necessary for teachers to teach in the 4IR.

In both the South African and the Swedish curricula there is an overarching goal that teaching, in all subjects, should shed light on sustainability issues. Learners should be given the opportunity both to take responsibility for the environment that directly influences themselves and to acquire a personal approach to global environmental issues. This includes responsibility and approaches to ways of living and working that can be adapted to create sustainable development. Thus, we can understand that there are sustainability issues that not necessarily are explicitly mentioned in the syllabuses, which have to be highlighted as well. The issue of sustainability is still vague even in the new curricula and in the teacher education programme.

### *Digital competence and ESD in science and technology teacher education*

The crucial question is, to what extent does teacher education in both institutions equip the student teachers for the demands of digital competence and ESD in schools, and what more needs to be done? In the South African university handbook, there is not much emphasis on the issues of digital competence and especially ESD and which leaves it to the choice of the lecturer who teaches the courses whether to integrate and equip the prospective teachers or not. The course outlines, therefore, are informed by an individual's willingness and expertise in digital competence. The Curriculum and Assessment Policy Statement (CAPS) document (Department of Basic Education, 2011) only provides student teachers and in-service teachers with topics, tasks as well as the duration of how topics need to be taught (Department of Basic Education, 2011). In Sweden, the emphasis on digital competence and ESD is based on what the head of the university is considering that teacher education should include, and what the teacher education institution as well as its teachers consider as necessary content in relation to the content of curricula in compulsory school. However, digital competence in the current teacher education is generally meaning the use of digital technology in various ways to improve teaching. The previous study at the University of Gothenburg by Sofkova Hashemi et al. (2019) shows that digital technologies are not well integrated into teacher education in practice, and Sofkova Hashemi et al. suggest that more must be done. In addition, the findings from the present study show that the syllabuses for technology and science education encompass content with respect to digital competence that go further than only the use of digital technology in teaching. For teacher education, which to a great extent is guided from content in the curricula of compulsory school, the findings in this study are important for further development of science and technology teacher education.

Based on the findings, we draw the conclusion that the curricula of compulsory schools have a rather broad coverage of digital competencies with respect to the DigComp 2.1 framework (Carretero et al., 2017). However, teacher education at the two institutions in the two countries does not yet meet this broad coverage when preparing student teachers. When implementing and broadening the coverage of digital competence, it will be especially important to incorporate ESD as an explicit and visualised part of digital competence. If digital technology will be the contributing force that will make it possible to achieve the SDGs, digital competence needs to be incorporated with sustainability-related content, together with teaching and learning approaches that stimulate and empower, both teachers and learners. This will provide digital competence that makes it possible to take action for sustainable development with the help of digital technologies. For being able to realise this, education needs to be transformed and employ new approaches (Leicht et al., 2018), in which teacher education plays an important role to provide pertinent knowledge and skills for future teachers.

Instefjord and Munthe (2016) argue that digital competence for teachers is a necessary complex set of skills and knowledge, not only for their own use of digital technology, but also for educating learners regarding skills and knowledge related to digital competence. For being able to provide this in teaching, they need to adopt compatibility to the different knowledge areas where digital competence is required (Instefjord & Munthe, 2016), as in science and technology education. Hence, it is important for teacher education to provide a content, which meets the requirements of the digital competence that teachers need (Krumsvik, 2011). The findings of this study show that when implementing digitals competence in science and technology teacher education, it is necessary to incorporate sustainability issues related to critical reflection on digital technologies regarding their potential, challenges and consequences on human, society and the environment. Further, it is necessary to direct attention to the ways digital technologies may be a contributing force to achieve the SDGs as well as what this requires in terms of digital knowledge and skills, values, attitudes and action competencies. Thus, as Leicht et al. (2018) argue: it is necessary to employ new approaches where human well-being and global development should be in focus with the aim to educate individuals as agents of change.

Considering the challenges in science and technology teacher education regarding digital competence from the perspectives of the SDGs and ESD in two different institutions, in South Africa and Sweden, we still need to be more digitally equipped. Teacher trainers need knowledge and skills concerning digital competence and ESD needs to improve to adequately prepare student teachers for the 4IR in the Anthropocene.

## Conclusion

Despite differences between countries across the world regarding prerequisites and access to digital technology and digital infrastructure, teacher educations across the world are facing similar demands in school curricula of providing learners with digital competence. This study shows that besides different prerequisites, and the different terminologies and use of vocabulary in the South African and Swedish education systems, the way the two countries embrace the move from traditional education to education acknowledging the digital age, happens in the same context. The two countries' education systems are clearly moving into the direction that the world is moving. Teacher training institutions ought to come on board and align their curricula in a way that is responsive to the needs of the community that they serve. Digital tools, connectivity, proper training of personnel as well as changing the education classroom and laboratories need to speak to the 4IR's vision as well as to the vision of digital technology as a contributing force for achieving the SDGs. To this end we both need to produce a digitally competent community of science and technology teachers that stimulate and educate individuals as responsible agents of change. This requires a teacher education that equips student teachers with digital skills and competencies to deliver digital competence incorporated with ESD in schools. This study points towards a gap in today's science and technology teacher education that is necessary for stakeholders to be aware of. Science and technology teacher education may contribute with its broad trans-disciplinary perspective on technology, society, humans and nature, but more is to be done. Further research is necessary for defining sustainable teacher practices promoted by questions on what to teach, and how to teach, regarding digital technologies and ESD.

## References

Atchoarena, D., Selwyn, N., Chakroun, B., Miaho, F., West, M., & Coligny, C. D. (2017). *Working group on education: Digital skills for life and work*. UNESCO. https://unesdoc.unesco.org/ark:/48223/pf0000259013

Berkout, F., & Hertin, J. (2004). De-materialising and re-materialising: Digital technologies and the environment. *Futures*, *36*(8), 903–920.

Carretero, S., Vuorikari, R., & Punie, Y. (2017). *DigComp 2.1: The Digital Competence Framework for Citizens with Eight Proficiency Levels and Examples of Use*. EUR 28558 EN. doi: 10.2760/38842

Department of Basic Education. (2011). *Curriculum and Assessment Policy Statement – Technology Grades 7–9*. https://wcedeportal.co.za/eresource/112886

Department of Higher Education and Training (DHET). (2015). *National qualifications framework, act 67 of 2008. Policy on the minimum requirements for teacher education qualifications*. Pretoria: Government Printers.

Digitaliseringskommissionen. (2014). *En digital agenda i människans tjänst: en ljusnande framtid kan bli vår : delbetänkande*. Stockholm: Fritze.

Donnelly, K. (2007). Australia's adoption of outcomes based education: A critique. *Issues in Educational Research*, *17*(2), 183–206.

du Plessis, E. C., & Marais, P. (2015). Emotional experiences of student teachers during teaching practice : A case study at Unisa. *Progressio*, *35*(1), 206–221.

Ferrari, A., Punie, Y., & Redecker, C. (2012). Understanding digital competence in the 21st century: An analysis of current frameworks. In A. Ravenscroft, S. Lindstaedt, C. D. Kloos, & D. Hernández-Leo (Eds.), *21st century learning for 21st century skills*. EC-TEL 2012. Lecture Notes in Computer Science, Vol. 7563. Berlin, Heidelberg: Springer.

Global e-Sustainability Initiative. (2017). *Summary Report. #SystemTransformation. How digital solutions will drive progress towards the sustainable development goals*. Accenture AG. Accessed 02/11/2019 from: http://systemtransformation-sdg.gesi.org/160608_GeSI_SystemTransformation.pdf

Ilomäki, L., Paavola, S., Lakkala, M., & Kantosalo, A. (2016). Digital competence – An emergent boundary concept for policy and educational research. *Education and Information Technologies*, *21*(3), 655–679. doi: 10.1007/s10639-014-9346-4

Instefjord, E., & Munthe, E. (2016). Preparing pre-service teachers to integrate technology: An analysis of the emphasis on digital competence in teacher education curricula. *European Journal of Teacher Education*, *39*(1), 77–93.

Krumsvik, R. J. (2011). Digital competence in the Norwegian teacher education and schools. *Högre Utbildning*, *1*(1), 39–51. https://hogreutbildning.se/index.php/hu/article/view/874

Labuscagne, C. (2018). *DBE Concept Document: Introducing Digital Skills for All into GET*. https://wcedeportal.co.za/eresource/141116

Leicht, A., Heiss, J., & Byun, W. J. (2018). *Issues and trends in education for sustainable development* (Vol. 5). Paris: UNESCO Publishing.

Penprase B.E. (2018). The Fourth Industrial Revolution and Higher Education. In: Gleason N. (eds) *Higher Education in the Era of the Fourth Industrial Revolution*. Singapore: Palgrave Macmillan. doi: 10.1007/978-981-13-0194-0_9

Redecker, C. (2017). *European Framework for the Digital Competence of Educators: DigCompEdu*. JRC Working Papers JRC107466, Joint Research Centre (Seville site). http://publications.jrc.ec.europa.eu/repository/handle/JRC107466

Rieckmann, M. (2018). Learning to transform the world: Key competencies in education for sustainable development. In A. Leicht, J. Heiss & W. J. Byun (Eds), *Issues and Trends in Education for Sustainable Development*, 39–59. Paris: UNESCO Publishing.

Schwab, K. (2017). *The fourth industrial revolution*. New York, NY: Crown Business Books.

Shulman, L. (1987). Knowledge and teaching: Foundations of the new reform. *Harvard Educational Review*, *57*(1), 1–22.

Skolverket (2017). *Få syn på digitaliseringen på grundskolenivå: ett kommentar-material till läroplanerna för förskoleklass, fritidshem och grundskoleutbildning*. Stockholm: Skolverket Wolters Kluwer.

Skolverket (2018). *Curriculum for the compulsory school, preschool class and school-age educare 2011: Revised 2018*. Stockholm: Skolverket.

Sofkova Hashemi, S., Player-Koro, C., Brante, E. W., von Otter, A.-M., Pihl, A.-L., & Franck, O. (2019). Lärarutbildningen och den digitala kompetensen. In O. Franck (Ed.), *Vetenskaplighet i högre utbildning: erfarenheter från lärarutbildning* (1st ed.). Lund: Studentlitteratur.

Taylor, N., & Vinjevold, P. (Eds.). (1999). *Getting learning right: Report of the President's Education Initiative Research Project*. Joint Education Trust.

United Nations. (2019). *The Impact of Digital Technologies*. https://www.un.org/en/un75/impact-digital-technologies

# 10 Science and technology teacher education in a global world

## Concluding remarks

*Elaosi Vhurumuku, Miranda Rocksén, Maria Svensson, Audrey Msimanga and Emmanuel Mushayikwa*

### Introduction

The challenges of the Anthropocene call for a widening of knowledge perspectives in science and technology teacher education. As said in our introductory chapter, this book is a result from a research collaboration project that brought together science and technology education researchers and practitioners from universities in South Africa and Sweden through workshops focusing on comparing theoretical perspectives, experiences and practices, and sharing research findings in response to anthropogenic challenges. Grounded in comparative and international studies in education and located within the realm of science and technology education for sustainable development (STESD), the topics covered in the book's ten chapters provide perceptive contributions to the understanding and knowledge of what sustainability issues might mean in the North (Sweden) and in the South (South Africa). The findings of the studies reported in this book offer instructive intuitions regarding how science and technology teachers might be prepared for work in their future profession and make contributions towards tackling the problems and challenges faced by their communities and the global world, in this era of the Anthropocene.

In this concluding chapter, first, we summarise the main findings coming out from the studies in the book. This is done through the reflective mirror of the four major aims of the book, which cover: a critical appraisal of the theoretical perspectives that have informed and guided science and technology teacher education in the Anthropocene; a comparison of policies and curricula in South Africa and Sweden; a comparison of experiences and pedagogical practices; and an exploration of opportunities and possibilities for fulfilling the goals of STESD. Within this effort, we present a comparative analysis of ESD in teacher education programmes at the participating institutions in South Africa and Sweden. Our comparison fleshes out the major similarities and differences in theoretical perspectives, policies, curricula, experiences

DOI: 10.4324/9781003190158-13

and pedagogical practices. Second, deriving from the experiences shared in the book, we take a brief foray into unpacking the meaning of local, regional and global challenges of sustainability. Third, we interrogate our own perspective of the Anthropocene and tease out some of the future challenges for science and technology teacher programmes. In our conclusion, we provide insights for future research and practice in science and technology teacher education.

## A summary of the major findings in the book

In this book inquiries were made in relation to four aims, and the individual chapters have reported results. A critical analysis of the theoretical perspectives underpinning science and technology teacher education revealed that, in both countries, teacher education is influenced by a wide array of theories and models including cognitive theories, critical theory, socio-ecological theory, theories on systems thinking and pedagogical content knowledge (PCK). In many instances these theoretical perspectives are spiced with ideas about globality. While there are these commonalities, there are also identifiable differences in tastes for the theoretical underpinnings in both ESD and the practice of science and technology teacher education. As can be inferred from *Chapter 1* of the book, in Sweden, the theory and practice in science and technology teacher education appears to be increasingly under the influence of systems thinking theories and perspectives. In South Africa, however, the theory and practice of STESD is largely under the influence of PCK models, transformation agendas, indigenisation and decolonisation of education. An interesting differential is that in South Africa, there is agitation for the integration of indigenous knowledge (IK) and indigenous knowledge systems (IKS) into mainstream science education including issues of sustainable development. In Sweden, the major focus appears to be the development of the sustainability literacy of teachers and students. As observed in Chapter 1, this differential is unsurprising, given that in the North (Sweden) science and technology teacher education is attuned to neo-liberal and Western imperatives and goals, whereas, for the South (South Africa), the theory and practice of science and technology teacher education appears to be driven more by socio-political voices calling for decolonisation of education and the indigenisation of policies and curricula to achieve the aspirations of the post-apartheid era.

*Chapter 2* of the book compares policies and curricula in South Africa and Sweden. This comparison shows that the policies and curricula of the two countries point at similarities and differences. In the comparison, the authors analysed curricula for the following levels; compulsory schooling, the upper secondary or Further Education and Training (FET) and university teacher education. They also analysed legislative and policy documents, guiding teaching and learning at all levels of the education system. Their analyses sought to determine how the curriculum documents addressed issues of sustainability, the ways in which legislative and policy guidelines promoted the teaching of science and technology for sustainability and how South Africa and Sweden compared in addressing issues in the Anthropocene.

Overall, for both countries, the analysis reveals a policy and practice dissonance. In other words, there is a lack of alignment between policy and practice. They also observe that, for both countries, issues of sustainability or ESD are not explicitly addressed neither in policy nor in curricula documents. Instead, issues of sustainability are mainly implicitly addressed within specific topics in the different subjects. They also found that, in Sweden, the aims in upper secondary schooling differ from compulsory schooling, whereas the aims do not change across the school levels for South Africa. The term Anthropocene is not mentioned in any of the policy or curricula in the two countries; however, in both countries, curricula aim at highlighting the importance of respect for the environment. Further, the chapter authors observe that in Sweden teachers are given a significant latitude to plan for and teach about ESD, whereas, in South Africa, teachers are constrained by the curriculum and don't have the leeway to plan and teach what they think is important in ESD. The conclusion is that, for both countries, policy and curricula documents are not sufficiently articulate about what are the learning goals to be achieved. Our argument is that if the long-term goals of science and technology teacher education for sustainable development are to be achieved, it is important that at all levels across the education ladder, ESD must be planned for, implemented and assessed.

*Chapter 3* argues that a back to the roots shift needs to happen in education, for us to be able to change our relationship with the environment and cultivate sustainable symbiotic relationships. The authors claim that it is now necessary to ask the big and existential questions such as who are we? what are we here for? where do we come from? how big is space? and how do we know what we know? Further, they propose that humanity has entered a new stage of evolution: epistemic evolution. Just as cultural evolution emerged against the background of biological evolution, epistemic evolution began as an aspect of cultural evolution. Epistemic evolution is characterised by the increasing dependence of global society on the achievements and further extension of science and technology in order to ensure its sustainability in the age of the Anthropocene.

In *Chapter 4*, a claim is made of multilingualism being a global phenomenon present not only in classrooms in the two countries, but in every classroom. Examples from classrooms in both countries illustrate how teachers and students move between languages, between English and Swedish; English and *isiXhosa* isiZulu; on the continuum between public language, which is considered the disciplinary language of chemistry and biology, and private language, used between the students but also by the teacher when providing examples. As a consequence, the chapter authors emphasise that teacher education needs to prepare student teachers for the multilingual classrooms.

In *Chapter 5*, the authors discuss the culture of recycling and reuse as a springboard for sustainability, something that technology teachers should emphasise. Environmental issues are part of the design process where technology solutions are developed in technology education. After comparing teaching of the design process in the two countries they describe cradle-to-cradle as a new phenomenon that ought to be embraced, including as a means of energy renewal. The chapter authors conclude that these two concepts should be promoted in the technology classroom in most of the activities given to students.

In *Chapter 6*, three themes emerge from professional reflective conversations between student teachers and teacher educators. These are: the need for student teachers to (1) have holistic understandings of ESD and STEM teaching; (2) learn about ESD in the STEM-subjects and how it is relevant to everyday life; and (3) understand how to model students' sustainability attitudes and behaviours. Underlying these themes is the recognition that ESD is complex and there is an increasing need for teachers to integrate content from different areas. The authors contend that, in both countries, outside of the school practice periods there are few official opportunities in teacher education programmes for structured reflective conversations. They posit that the value of professional reflective conversations may be underestimated.

*Chapter 7* addresses the challenges of how to prepare prospective Biology teachers to implement and integrate ESD in the classroom. In general, the authors found that in both countries, policies guide the initial preparation of Biology teachers for ESD in a very implicit way with a strong bias towards the environmental dimensions of ESD. In the absence of explicit policy guidelines, the initial preparation of Biology teachers for ESD heavily depends on teacher educators' agencies, beliefs and perceptions developed from previous work and learning experiences in varied sustainable development areas. The chapter authors make a recommendation that ESD should be systematically integrated across all the courses that student teachers take during their preparation, including teaching practice. This integration provides an opportunity for curriculum developers to engage in curriculum decolonisation processes aimed at knowledge generation by local players and framed around ESD.

The concept of culturally embodied learning was explored in *Chapter 8*. This was done through an evaluation of two interventions done in South Africa and in Sweden. In both cases, cultural knowledge and artefacts were embedded in the teaching of scientific concepts. The chapter illustrates the possibility of integrating specific culture with Western science. This affords communities opportunities to adopt scientific reasoning into their cultural practices. An analysis of the merits of the two interventions suggests that culturally embodied learning is suitable as an interpretive frame that enables discussions and deeper understanding of challenges associated with multicultural science education. The authors concluded making science relevant to both the student and his/her environment using culturally embodied learning can be useful in the development of multicultural science teacher education. They argue that this is useful in addressing issues of sustainability in the contemporary era.

In *Chapter 9*, it is shown that besides different prerequisites, and the different terminologies and use of vocabulary in the South African and Swedish education systems, the way the two countries embrace the move from traditional education to education acknowledging the digital age happens in similar contexts. The authors conclude that it is important to incorporate ESD as an explicit and visualised part of digital competence. If digital technology is to be the contributing force that will make it possible to achieve sustainable development goals (SDGs), then digital competence needs to be incorporated with sustainability-related content, together with teaching and learning approaches that stimulate and empower both teachers and learners.

Drawing from the findings from the book chapters, we argue that the challenges of the Anthropocene are complex and immense and call for a widening of knowledge perspectives in science and technology education. Two observations emerge from the findings reported in this book: Firstly, although the two countries are located in different geographical and socio-economic regions, and even though they are ranked differently on the Human Development Index, they face similar ESD challenges in STEM education. Secondly, the two countries have different ways of dealing with the challenges.

## Connecting the local, regional and global

Throughout history, educational systems have continuously been transformed. In South Africa significant transformation was triggered by the abolition of apartheid and the development of one system of education for all students during the 1990s. For Sweden, the comprehensive school was introduced during the 1960s as a consequence and in the aftermath of World War II. For both countries, however, the legacy of is evident. This is illustrated by the structure of the school system and the use of English as the main language of instruction in classrooms throughout South Africa for example. In Sweden, the history of exploitation of the Sàmi people continues in current debates about education, and Sàmi language instruction is only provided in a small number of schools.

Globalisation does not have the same significance and consequences for all countries. One reason for using these two perspectives – South and North – is that international comparative studies enable contingency. It is a platform for investigating the interplay between, in our case, educational phenomena and their different contexts on the local, regional and global levels. The answers to questions such as *How do we educate our children? What does students' and teachers' education look like?* vary depending on at what place they are asked and at what time. Comparisons between different times and places can make us aware of our own educational traditions, and the choices we have made or not made. The changes that are taking place in education around the world are often linked to each other. New ideas are circulated and disseminated by various actors with the purpose to create change. These actors can be international organisations (e.g., the UNESCO and the OECD), universities, researchers, politicians and other decision-makers. According to The Global Sustainable Development Report (2019), neither South Africa nor Sweden represents a place where basic human needs and sustainable resource use are in balance. This report shows South Africa and Sweden as two of the countries where biophysical thresholds are exceeded. Albeit, Sweden's values are higher than South Africa's. However, it is on the scale that indicates the extent to which social thresholds are reached, that the difference between the countries is most clearly seen. According to the report, South Africa reaches fewer social thresholds than Sweden and the countries are far apart on the scale. We believe that a North and South perspective more generally opens opportunities to critically reflect on how science and technology teacher education prepares students for current changes in society and the world and prepares new teachers appropriately.

## A notion of the Anthropocene

The challenges of Anthropocene call for a widening of knowledge perspectives in science and technology education. Climate change, biodiversity loss and the COVID-19 pandemic represent three crises affecting the circumstances for living organisms in their respective local environments around the world. In order to support students' understanding of the effects of a crisis like climate change, teachers in Chemistry, Technology, Biology and Physics bring in explanations including concepts like atmospheric gaseous composition, effects on biodiversity and energy. In this concluding chapter we would argue that the challenges of today are even more complex and immense and call for a widening of knowledge perspectives in science and technology teacher education. For us as science and technology teacher educators we find ourselves and the task of teaching becoming teachers as being within what Timothy Morton (2012, 2013) calls a hyperobject.

The notion of hyperobject is coined by Timothy Morton (2012, 2013). Morton approaches Anthropocene from the perspective of literary art and philosophy. In his description Anthropocene reflects the asymmetry between human capacity for knowledge and gigantic hyperobjects: "...entities that are massively distributed in time and space" (Morton, 2013, p. 39). According to Morton, entities of this scale, like climate change, biosphere, nuclear radiation, affect everything, cannot be localised and occupy the space of the manifestation of a system. In this way the notion of hyperobjects shifts the perspective on human knowledge. It is a shift going from seeing human scientific knowledge as the measurement of its local manifestations to the realisation of being inside this immense entity.

What does this perspective mean for becoming a teacher in the science and technology? Viewing the role of science and technology teachers from the perspective of them being themselves inside a hyperobject, being directed to teach about measurements and subject specific concepts, we think works as an illustration of the difficult endeavour to seriously include issues of sustainability in the teaching of science and technology. There are few tools, limited language, concepts and approaches for including holistic perspectives or more locally developed knowledge. Instead, we find that educational systems are increasingly compared and streamlined through international testing, neglecting perspectives which are local, place specific, and have emerged in relation to communities and cultures at a certain geographical location.

## Future challenges for science and technology teacher education

Science and technology teacher education programmes play an important role in creating an understanding and preparation for dealing with the increasingly urgent issues and problems of sustainable development. Education of science and technology teachers cannot alone change our understanding of problems, but is one part, which can make a contribution. This role is emphasised in the post-truth era in which scientific knowledge has been contested in new ways. Education can help people cope in a complex and demanding environment where cross-curricular knowledge becomes essential. One challenge for science and technology teacher education might be to change from well-established approaches of teaching to new and less established ones. Feinstein and Waddington (2020) describe two directions of science education, the internalist approach and the contextual approach, and they stress that contextual and sociocultural approaches to science education offer a better tool for responding to the post-truth era. The internalist approach focuses on making a stronger case for the specific knowledge of scientific practices and the robust nature of scientific findings, while the contextual approach focuses on the relationship between science and other knowledge areas as well as between science and issues in everyday life. This is in line with the way Roberts (2007) describes science education as having two visions, Vision I and Vision II. Where Vision I "looks inward at science itself – its products such as laws and theories, and its processes such as hypothesizing and experimenting", whereas Vision II "looks outward at situations in which science has a role, such as decision-making about socioscientific issues" (Roberts, 2007, p. 9). Sustainable development cannot be limited to an internalist approach or to Vision I, but requires understanding of socially significant issues on pragmatic grounds. The idea of understanding socially significant issues on pragmatic grounds is also supported by Brennan (2017), who argues that, because the challenges faced by humanity in the Anthropocene are intricately intertwined with the crises of capitalism, it is necessary for teacher education to adopt pragmatist philosophy and rethink "... the tasks which teacher education can contribute to an activist-oriented development of education in these difficult times for humans and the planet" (Brennan, 2017, p. 43). Therefore, in this post-truth era, we believe that science and technology teacher education needs to put sustainability in the foreground, integrate Western scientific knowledge with indigenous knowledge and work in a cross-curricular manner with cross-cultural aspects. This would mean not only to implement sustainability issues as topics in science and technology education, but also to develop sustainable teaching strategies, leading to a more sustainable science and technology education. In this perspective, science and technology teacher education is one part, with a significant role to play in order to prepare becoming teachers to understand situated teaching practices and being able to adjust their teaching to learners from different cultural settings.

## Conclusion

It is not enough to just integrate science and technology education for sustainable development into the curricula of teacher education programmes, but also to ensure that the teachers have adequate knowledge, competencies and skills, enabling them to develop learners into citizens who understand and appreciate the environmental, social, economic, ecological, ethical and political aspects of sustainability. This is a much more demanding challenge and requires resources and engagement at all levels of education. Political decisions regulate and influence the debate about sustainability and is something that student teachers must be aware about and understand, to be able to handle in the teaching of science and technology. This points to the importance of developing research-informed practices in teacher education, including for example the teaching and learning of subject-specific teacher competence, breadth in instructional strategies and how to address sustainable development issues, as well as providing students methods for developing their own teaching as professionals. These practices can be achieved by means of international research, collaboration and comparison.

Kwatizhe (2015) asserts that assessment is a critical determinant of curriculum implementation. This is especially so for developing countries (see, for example, Mkandawire, 2010; Mufanechiya, 2012). Taking the challenges of the Anthropocene seriously would mean also to articulate some basic principles, models and concepts that ought to be explicitly included in curricula. Only with a more precise discourse on what is expected from teachers in their teaching in terms of ESD, can the teachers adjust to anthropogenic challenges. It is only with a more precise discourse on what is expected from teachers in their teaching that they will be able to seriously take the task of teaching ESD explicitly in planning, curriculum implementation and assessment.

Our recommendation is that for both the North and the South, policies and curricula must make explicit the content for STESD rather than leaving this implicit. We suggest for future research in STESD to investigate such issues as comparing the sustainability competences of students and teachers in the North and the South, curriculum models and pedagogical practices, which promote teachers' and learners' competences in sustainability. This, in our view, will go a long way towards achieving the goals of global sustainable development.

## References

Brennan, M. (2017). Struggles for teacher education in the age of the Anthropocene. *Journal of Education (University of KwaZulu-Natal)*, (69), 43–66.

Feinstein, N. W., & Waddington, D. I. (2020). Individual truth judgments or purposeful, collective sensemaking? Rethinking science education's response to the post-truth era. *Educational Psychologist, 55*(3), 155–166.

Kwatizhe, S. (2015). Factors affecting curriculum implementation for students. *International Journal of Advanced Research, 1*(12), 984–986.

Mkandawire, S. B. (2010). Impediments to curriculum implementation in learning institutions. *African Higher Education Review, 8*(2), 1–15.

Morton, T. (2012). From modernity to the Anthropocene: Ecology and art in the age of asymmetry. *International Social Science Journal, 63*(207–208), 39–51.

Morton, T. (2013). *Hyperobjects: Philosophy and ecology after the end of the world.* Minneapolis, MN: University of Minnesota Press.

Mufanechiya, T. (2012). The dog wagging the tail or the tail wagging the dog? The impact of national examinations on curriculum implementation in Zimbabwean secondary schools. *Journal of Emerging Trends in Educational Research and Policy Studies, 3*(5), 655–660.

Roberts, D. (2007). *Scientific literacy/science literacy.* In S. Abell, & N. Ledermann (Eds.), *Handbook of research on science education* (pp. 729–780). Mahwah, NJ/London: Lawrence Erlbaum Association.

United Nations (Independent Group of Scientists appointed by the Secretary-General). (2019). *Global sustainable development report 2019: The future is now – science for achieving sustainable development.* New York, NY: United Nations.

# Index

Pages in *italics* refer to figures, and bold refer to tables

For Product Safety Concerns and Information please contact our EU
representative GPSR@taylorandfrancis.com
Taylor & Francis Verlag GmbH, Kaufingerstraße 24, 80331 München, Germany

www.ingramcontent.com/pod-product-compliance
Lightning Source LLC
LaVergne TN
LVHW021126300926
849160LV00010B/883
*9781032039978*